Let's Go
Sociology

Travels On The Internet

Second Edition

Joan Ferrante and Angela Vaughn
Northern Kentucky University

Wadsworth Publishing Company
I(T)P® An International Thomson Publishing Company

Belmont, CA • Albany, NY • Boston • Cincinnati • Johannesburg • London • Madrid • Melbourne
Mexico City • New York • Pacific Grove, CA • Scottsdale, AZ • Singapore • Tokyo • Toronto

For more information, contact Wadsworth Publishing Company, 10 Davis Drive, Belmont, CA 94002, or electronically at http://www.wadsworth.com.

International Thomson Publishing Europe
Berkshire House
168-173 High Holborn
London, WC1V 7AA, United Kingdom

International Thomson Editores
Seneca, 53
Colonia Polanco
11560 México D.F. México

Nelson ITP, Australia
102 Dodds Street
South Melbourne
Victoria 3205 Australia

International Thomson Publishing Asia
60 Albert Street #15-01
Albert Complex
Singapore 189969

Nelson Canada
1120 Birchmount Road
Scarborough, Ontario
Canada M1K 5G4

International Thomson Publishing Southern Africa
Building 18, Constantia Square
138 Sixteenth Road, P.O. Box 2459
Halfway House, 1685 South Africa

International Thomson Publishing Japan
Hirakawa-cho Kyowa Building, 3F
2-2-1 Hirakawa-cho
Chiyoda-ku
Tokyo 102, Japan

ISBN 0-534-53666-2

To
Harold and Linda Vaughn

Table of Contents

Preface
from
Joan Ferrante

The first edition of *Let's Go Sociology: Travels on the Internet* was released on January 2, 1997. The second edition was submitted for publication on May 27, 1998. The bad news is that between those dates, 37 percent, or 225 out of 613, of the Web sites changed. The good news is that with the exception of approximately 10 sites, all the new Web sites could be easily found with the help of a search engine and by using the document name as key word. The point is that while *Let's Go Sociology: Travels on the Internet* is a useful guide to the kinds of information that can be found on the Internet, you should expect the addresses to change. You should also feel confident that the new address can be found simply by using a search engine. Both editions of *Let's Go Sociology: Travels on the Internet* were created with the college student in mind, especially sociology students. We strived to create a resource guide with URLs that would enable students to do the background research needed to complete most assignments from the comfort of their homes or dorm rooms. In addition, we strived to create a book with URLs containing important information that sociology instructors can assign students to read in conjunction with other, more traditional, reading materials. Instructors may also wish to design assignments around the information resources included in this book.

The pedagogical significance of *Let's Go Sociology* is especially relevant in light of the growing popularity and widespread access to the World Wide Web, an information-sharing tool that undermines central control over information and fosters connectedness and cooperation. In "Redefining Success: Public Education in the 21st Century," Yvonne Katz and Gay Chedester (1992) point out that as more and more students come to school "with notebook-size computers tapped into all the world's knowledge," the instructor's major task will be to teach the science of choice (the ability to evaluate and choose from a wide variety of resources) and the importance of interconnectedness (ability to communicate, collaborate, and cooperate in creating information). The new emphasis will push educators and publishers away from viewing students as passive consumers of information created by others to viewing them as evaluators, synthesizers, and contributors of information (Crawford 1995).

The most enjoyable part of this project was collaborating with Angela Vaughn—a capable, disciplined, and inspiring student—to create a resource directory *for students*. I met Angela in the spring of 1996 when she was a student in my Introduction to Sociology class. There were 85 students, a number far too large for me to know everyone on a first-name basis. As the semester drew to a close, I decided to look down the list of names in my grade book with the hope of identifying a good and reliable student whom I could approach about working as my research assistant for the summer. Although there were many good candidates to choose from, one student's record stood out.. On the next to last day of class, I asked, " Is Angela Vaughn here?" Angela raised her hand. I asked if she would please see me after class assuring her that she was not in any kind of trouble.

Angela accepted my employment offer and began helping me create an Internet home library for an introductory sociology book I was working on *(sociology.net: Sociology on the Internet)*. In the matter of a few weeks it was clear to me that we worked well together and that Angela possessed the conceptual skills needed to sift through, organize, comprehend, and evaluate the staggering and ever-increasing amount of information posted on the Internet. We decided to expand the home library and create *Let's Go Sociology: Travels on the Internet.*

We dedicate this book to Angela's parents, Harold and Linda Vaughn, two people who make my job as instructor easy, as they nurture, support, and encourage academic excellence.

Reference

Crawford, Jack. 1995. " Renaissance Two: Second Coming of the
    Printing Press?" http://www.lincoln.ac.nz/reg/futures/renaiss2.htm

Katz, Yvonne, Gay Chedester. 1992. " Redefining Success: Public Education
    in the 21st Century." *Community Services CATALYST 22(3)*
    gopher://borg.lib.vt.edu/00/catalyst/v22n3/katz.v22n3

Preface
from
Angela Vaughn

In the spring of 1996 I enrolled in an Introduction to Sociology class taught by Dr. Joan Ferrante, a professor I did not know. In May 1996 I accepted an offer to be her research assistant. I began working on the Internet right away, though not with this project in mind. Although I was familiar with computers before the summer of 1996, I had never been on the Internet. In less than 30 minutes Dr. Ferrante taught me the basic skills I needed to navigate the Internet. Initially my job was to search for sites of value to students and to make a "home library" for a textbook Dr. Ferrante was working on about the Internet. It soon became obvious to both of us that this directory of Web sites could be expanded and become its own book. After several weeks, Dr. Ferrante gave me the opportunity to work with her as co-author on this expanded book.

As a student, I find the Internet to be a great resource for research papers and projects. *Let's Go Sociology: Travels on the Internet* is a valuable tool for students because it saves time searching for information and it offers concise summaries of the kinds of information posted at each site. I find myself referring to this book whenever I need information quickly, and this book is useful even for assignments outside the discipline of sociology. I hope that other students using this guide will find their travels on the Internet less frustrating than if they were conducting a search on their own.

It would be impossible to thank everyone who contributed to this project, because many of the most important contributions were made by dedicated individuals and institutions from around the world who took the time to support or post high-quality information on the Web. Dr. Ferrante and I have decided to donate a portion of the royalties we receive on this project to support and promote access to the Internet.

My deepest thanks goes to Dr. Ferrante for giving me the opportunity to work with her on this book and asking me to be a co-author.I found working one-on-one with a professor to be the most valuable learning experience that resulted from this project. I thank her for her constructive criticism and for her patience with my inexperience.

I dedicate this book to my parents, Harold and Linda Vaughn, for their excellent example, unwavering support, and unconditional love.

Introduction to the Internet

Let's Go Sociology: Travels on the Internet is not a book about the Internet. There are already many such available for those who wish to learn the Internet's extensive vocabulary and to acquire skills that go beyond those needed here. Consequently we will cover only what you need to know to access the Internet sites included in this book. Specifically you need to know the answers to these questions: (1) What is the Internet? (2) What is the World Wide Web? (3) What is a browser? (4) What are Uniform Resource Locators (URL's)? (5) What is a search engine? and (6) What is a local Internet access provider? (or How can I gain access to the World Wide Web?)

What Is the Internet?

The Internet is a vast network of computer networks. A computer network is a system of computers connected to one another via special software and phone, fiber-optics, or other types of lines. The Internet links together tens of thousands of in-house computer networks maintained by businesses, libraries, government agencies, universities, and private organizations.

What Is the World Wide Web?

The *World Wide Web* is one of several Internet-based services (others include e-mail and Usenet, a discussion/newsgroup service). The Web which was invented in 1991 is a constantly changing and ever-expanding information-sharing tool that facilitates the exchange of text-, video-, and audio-based information. Because no one owns the Web or manages its content, it is impossible to calculate the number of computer networks involved or the amount of information available for exchange. According to one estimate some 500,000-and-counting documents are available through the Web (Steinberg 1996). Fortunately most of the documents available for access are formatted according to standard specification known as HTML (hypertext markup language). *Hypertext* connects documents to one another and allows users to move quickly via links within and across documents located anywhere on the Web. *Links* are highlighted words or images that set them apart from the rest of a document's text. After users choose a link or click a link with a mouse, they move elsewhere. Any browser can read any basic HTML document, unless the HTML format has special features that can be read only by particular browsers.

What is a browser?

Users access information on the World Wide Web with a browser. There are two kind of browsers: character-based client browsers and graphical client browsers. *Character-based browsers* read only letters, numbers, spaces, and nongraphical marks or signs. Users navigate or move from one document to another via arrow keys and the space bar.

1

The most popular character browser is Lynx. In contrast, graphical browsers can process characters, images, and sounds. Two popular graphical browsers are Netscape Navigator and Microsoft Internet Explorer. To gain access to a browser, you need a local Internet provider (see "What is a local Internet provider?").

What Is a URL?

Every document on the Internet has an address called a *Uniform Resource Locator.* Some examples of URLs include the following:

1. http://www.usatoday.com/news/nweird.htm
2. http://library.whitehouse.gov/PressReleases-plain.cgi
3. http://www.etext.org/Politics/Disability.Rag
4. http://stats.bls.gov/ocohome.htm
5. http://www.wri.org
6. http://www.soc.surrey.ac.uk/socresonline

The rules for URLs are very loose, but there is a basic structure for URLs. Most of the URLs listed in *Let's Go Sociology: Travels on the Internet* begin with http://; some begin with gopher:// and a very few begin with ftp://. "http" stands for hypertext transfer protocol. "ftp" stands for file transfer protocol. Protocol is a system of procedures to access information. "gopher" is the name of the University of Minnesota's mascot. The University of Minnesota is the home of this Internet navigational tool which facilitates the search for information.

The domain name system or the host computer name appears between the double slashes (//) and the first single slash mark (/). It is called a domain name system because it makes reference to several domains or areas within an in-house computer network. Each domain is separated by a period. Notice that most of the domain name systems listed in our six sample URLs end with a three-letter code — .org, .com, .gov, or .edu. This tells us that the host computer is an organization (.org), company (.com), government agency (.gov), or educational institution (.edu). Sometimes the domain name system ends with a two-letter code known as a country code. For instance in example 6 .uk stands for United Kingdom; other such codes include .us for the United States and .tr stands for Turkey.

To the left of these two-digit or three-digit codes are usually some clues about the identity of the organization, company, government agency, or educational institution. It's easy to see that the URL http://www.usatoday.com/news/ nwierd.htm refers to the company USAToday and the second URL http://library.whitehouse.gov/PressReleases-plain.cgi to the White House, a department within the U.S. government. The domain name system in example 3 (http://www.etext.org/Politics/Disability.Rag) gives no clues to the identity of the organization. However, "etext" suggests that the URL leads to an electronic text. The "bls" in example 4 (http://stats.bls.gov/ocohome.htm) refers to the Bureau of Labor Statistics, the "wri" in example 5 (http://www.wri.org) to the World

Resources Institute, and the "surrey" in example 6 (http://www.soc.surrey.ac.uk/socresonline) for the University of Surrey in the United Kingdom.

The information between the single slashes represents codes for the directories or hypertext paths one must follow to get to a particular document. Clues about the name or kind of document to which the URL takes you are to the far right of the URL address. If there are no paths, as in example 5 (http://ww.wri.org), the URL takes you to the host institution's opening page or home page. The documents for each of our six examples are as follows:

1. nweird.htm	Weird News Stories
2. PressReleases-plain.cgi	Press Releases
3. Disability Rag	(a magazine that focuses on disability)
4. osh.txt	Workplace Injuries and Illnesses Report
5. No paths specified in URL	
6. socresonline	Sociological Research Online (an online journal)

To gain access to these documents, you simply type in the URL at the appropriate place designated by the browser. In the case of the Lynx browser, press the letter "g" key and begin typing the URL in the designated space. In the case of Netscape Navigator, delete the URL appearing in the "Location" box and type the URL you wish to locate in its place.

How do you find URLs? URL addresses can be found in many different places. Like this book, there are URL "Yellow Pages" which are organized by subject with relevant URL addresses. Sometimes corporations and agencies will list their URL address in an advertisement. For example, at the end of the Lehrer-Newshour, PBS lists their URL which is http://www.pbs.org. Another way to find URLs is via search engines.

What Is a Search Engine?

A *search engine* allows users to submit a keyword or words to identify the topic on which they wish to find information. The search engine identifies Web sites and corresponding URLs that lead to information on that topic. Some names of search engines are InfoSeek, Yahoo, CUSI, Starting Point, Lycos, Webcrawler, and Excite. Because there is so much information on the Web and no central authority managing its contents, "organizing the Web is probably the hardest information science problem out there" (Yang, in Steinberg 1996, p. 109). Search engines try "to bring order out of chaos in a frantic quest for the ultimate index of human knowledge" (Steinberg 1996, p. 109). Yahoo indexers, for example, claim to have catalogued more than 200,000 Web sites (and counting) into 20,000 different categories

(Steinberg 1996). To gain access to a search engine (and by extension the Internet), you need to find a local Internet access provider.

What Is a Local Internet Access Provider?

Most people can not afford their own direct connection to the Internet. Consequently they rely on a local Internet access provider, a nearby host institution that offers Internet access via a local telephone call as a benefit of membership or for a fee. Many students can obtain access to the Internet through their school or university. Some businesses and other organizations offer their employees or members access, and public libraries give access to patrons. In addition, dozens of commercial Internet service providers offer the connection for a fee (ranging from $10 dollars a month and up, depending on use). Examples include American Online, Prodigy, Compuserve, and perhaps your local telephone company.

To access the portion of the Internet known as the World Wide Web from home, you must have a personal computer, a modem, and modem software (don't panic), which allows you to connect to a provider's host computer via dial-up telephone lines. You will have to ask your local Internet service provider about software and other details for connecting to their service from your home.

If you are a student and do not have a personal computer and/or the necessary equipment to connect to the Internet from your home, check out the availability of public computer labs located in dorms, academic departments, libraries, and elsewhere on campus.

Some Words of Advice

We close with some advice about the Internet from some students at Northern Kentucky University who use the Internet. Their suggestions can save you from a lot of headaches. You might want to try some Internet sites before you read the advice.

Jenny DeBerry:

" The Internet presents a whole new way of thinking about information. My advice is to expect change and learn to be comfortable with it. People who post information on the Internet can choose to take it off or revise it whenever they see fit. With books, authors might want to change their ideas but have to wait until the next edition (if here is a next edition) to get it out there for the world to see. With the Internet, authors can revise their thinking whenever they see fit and make changes on the spot. So don't be surprised if you find that a document is gone or has changed in subtle or dramatic ways from one visit to the next."

Patricia Gaines:

" My advice is not to wait until the last minute to do Internet assignments. The Internet might not be available an hour before your assignment is due. I read an article in *The New York Times* that says it all: 'More than a million computer users suffered interrupted or erratic Internet and on-line service connections last week because of a variety of planned and unplanned service shutdowns' (Lewis 1996, p. C1). Sometimes the system crashes lasted as long as eight hours."

Ryan Huber:
　"My tip is to always double-check the URL you typed in against the one listed in the text. It must be typed *exactly* as listed in the text. For example, lowercase letters cannot be substituted for capital letters. This will prevent or solve many of the problems you might have in locating a site that you know is there but that the computer is telling you it isn't. Also, when using a search engine such as the Web Crawler, Excite, or Lycos, try to think of more than one key phrase or search word. I usually try to think of eight different keywords/phrases for each of the topics I am researching. Be patient and explore. You may not always find what you are looking for right away. Don't be afraid to follow hypertexted links; sometimes the best information is found in the most unusual places."

Laureen Norris:
　"Using the Internet requires you to have a great deal of patience, a low frustration level and a detectivelike mentality. To get the best results for a search on the Internet, be as specific as possible with your choice of words and phrases. The Internet has so many sites that if your search word is too vague or general, the search engine may generate tens of thousands of sites. I found that the hypertexted words or suggested links within documents often are more valuable to my research than the document the search engine generated as directly related. Sometimes the highlighted words or links appear to have no relationship with my chosen topic."

Julie Rack:
　"Sometimes the Internet can be very slow. You may have to wait 20 or 30 seconds (imagine that!) for the document to appear after you type in the URL. My advice is to try nonpeak-hour times before 9:00 a.m. and after 5:00 p.m. The fewer the people on the Internet, the faster the response time."

Jacob Stewart:
　"The amount of information on the Web is massive. You can find something on almost any topic. However, you must be mentally and physically tough when searching for information on the Web. Sometimes you can feel totally overwhelmed to the point of panic. But don't worry; most people feel this way at one time or another. My advice is to relax and to not feel that you have to look or read over everything on a topic."

Lindsay Hixson:

" I could list a hundred tips about the Internet, but I will offer just one. Many Web sites are rearranged or restructured on a regular basis so that the path one takes to get to a document might change. When that happens, the URL address listed in the textbook may not get you to the document. Don't be alarmed if you type in a URL and the message 'unable to connect to remote host' or 'a path does not exist' appears. The chances are very good that the document is still there but that the 'path' has changed. So what do you do? You could take the easy way out and tell your instructor that the URL didn't work, so you couldn't do the assignment. Or with a little effort you could try to find the document yourself. Here's what you do: Type in the entire URL and then erase the address back to the first slash. For example, if your URL is **http://www.cdinet.com/Rockefeller/Briefs/brief35.html**, erase **/brief35.html** and submit the revised address. If that revised address doesn't work, erase the address back to the next slash. In this case you will erase **/Briefs**. Keep doing this until you have only the domain name system to submit. If you gain access, search for hypertext links that match up with path code names in your original URL. Hopefully this will not happen too often, but if it does, you know what to do. Good luck."

Melissa Cox:

" When I started this project a friend suggested that I purchase a book covering the 'how to's' of the Internet. I replied, 'unless it is a book of URLs which lead to high-quality information, forget it.' The best way to learn about the Internet is to talk to others who are also using the Internet. It is surprising how much you can pick up on your own and by sharing tips with others. Also don't be afraid to ask questions if you get stuck."

References
Lewis, Peter H. 1996. " When On-Line Service Cannot Be Counted On."
The New York Times (June 24):C8.

Steinberg, Steve G. 1996. " Seek and You Shall Find (Maybe)."
Wired (May):108-115+.

Yang, Jerry. 1996. Quoted in " Seek and You Shall Find (Maybe)."
Wired (May):111.

Tips on Using
Let's Go Sociology: Travels on the Internet

There are many URL directories, known as "Internet yellow pages" on the market, some 600 to 700 pages long. Virtually all these directories try to be everything to everybody by including URLs related to almost every conceivable topic and interest (herbs, sports, on-line dating, jokes, gardening, games, home maintenance, skateboarding, oceanography, pornography, poetry, philosophy, and science fiction). In addition to offering a smorgasbord of information sources, the accompanying descriptions are often too general to help users evaluate the quality and kind of information posted on a site without going there first.

Let's Go Sociology: Travels on the Internet is a URL directory created with the college student in mind. It includes a list of URLs that take you to Web sites containing information that can help you do most college assignments, especially assignments related to sociology courses, without leaving your desk. (This will save wear and tear on you, your car, and the environment). There are over 500 URL addresses that will lead you to sites that provide a wide range of information from accepted abbreviations for international organizations and groups to zipcode-level data (more than 100 tables giving information on everything from age composition to the number of vacant housing units for a specified zipcode). Some URLs are cross-listed under more than one topic. However, in most cases URL descriptions are listed under only one of those topics. In such cases you are referred to the topic under which the description exists.

Although we have also tried to select Web sites that contain high-quality information and to write concise descriptions that specify the kinds of information you will find there, we suggest that, to get the most of this URL directory, you spend an hour or so reading and/or skimming the entries. First, this will help you appreciate the extraordinary capacity of the Internet to put a wide range of high quality information at your fingertips. Second, once you grasp the range of resources named in this book, you will be in a better position to respond to a variety of assignments (essays, term papers, research studies, poster, presentations, maps, and so on) in creative, flexible, and efficient ways. Your response can be *creative* because you have a range of thought-provoking and interesting resources from which to choose. You can be *flexible* because you can explore your resource options quickly. Finally you can be *efficient* because you can spend the time on your assignment that you will have spent driving to and from the library, tracking down sources, and standing in line to photocopy needed materials.

In addition we have a number of specific suggestions that will help in your search for the information you need:

*There are no separate entries for individual countries. If you are looking for information on a specific country, we suggest that you go to the Country-level section of this directory. Pay special attention to the documents "Country Background Notes" and the "World Factbook," both of which contain information on every country in the world.

*Rather than rely on newspapers and the network news sources for information, check out the "Press Releases" and "Briefings" section of this book. Much of the popular media's information and ideas for stories come in the form of press releases. Check out this section for first-hand accounts of information released to the press.

*If you are in the early stages of an assignment and you are simply looking for an interesting idea or an unusual angle on a well-known topic check out "Hot News/Hot Research" and the "Issues Page." Both of these sites are located under the "Social Issues" section of this directory and provide unusual and in-depth coverage on a number of common topics.

*If you are looking for information on specific racial or ethnic populations we suggest that you go to the U.S. Census bureau lookup site or use its search engine. The historical documents section contains Web sites which house some of the most influential writings, speeches, and legislative and legal actions related to civil rights. Also consider looking under headings related to the kind of information needed on a specific population. If you are interested in certain veteran groups or on employment discrimination check out the "Veterans" or "Labor" section.

*Descriptions of U.S. Government and international agencies are limited to the functions and goals of that organization. In most cases we do not describe the specific documents housed at these sites because there are simply too many to describe. The agencies give access to newsletters, schedules of major meetings, historical documents, statistics, press releases, and so on. These Web sites are important sources of information and can help narrow a general topic such as education, labor, or defense to more specific topics such as "alternatives to standardized tests," "projected job growth," and "health care needs of female veterans."

*If you would like to explore some Web sites in order to become acquainted with the Internet, we recommend the following five URLs:

- "Cultural Exchange" at http:// deil.lang.uiuc.edu/exchange/,
- "Emoticons" at http://www.organic.com/1800collect/emotions/index.html,
- "Movie Database" at http://us.imdb.com/tour.html,
- "Quotations Home Page" at http://www.lexmark.com/data/quote.html, and
- "Documents in the News" at http://www.lib.umich.edu/libhome/Documents.center

* Beginning on p.95 we have listed approximately 120 URLs that will take you to Web sites posting very specific kinds of information on a variety of topics.

* Please note that the letter "l" and the number "1" look very similar. If you have trouble accessing a Website, consider whether this similarity might be the source of the problem.

* Before this book was sent to press for the second edition we checked and updated all its URLs (approximately 500). We estimated that about 42 percent of the addresses had changed over an 18 month time period. If a URL does not work we recommend that you use the document name as key words and select a search engine to help you locate the Web site. If that does not help then we suggest that follow Lindsay Hixson's advise on page 5.

*Finally, be a responsible netizen, or citizen of the Internet. Read "Declarations of the Rights [and responsibilities] of Netizens" and "PolitnessMan's Guide to Netiquette." Take the time to e-mail people and organizations to thank them (or their sponsors) if you find useful the information they have posted.

Abbreviations

- **Abbreviations for International Organizations and Groups**
  http://www.odci.gov/cia/publications/factbook/app-frame.html
  Do you need to know what a particular abbreviation stands for? Or would you like to know the abbreviation of an organization for a project or paper? This site lists accepted abbreviations for international organizations and groups starting with the Arab Bank for Economic Development in Africa (ABEDA) and ending with the Zangger Committee (ZC).

Aging

- **Administration on Aging**
  http://www.aoa.dhhs.gov
  The U.S. Administration on Aging "has developed a nationwide network of State and Area Agencies on Aging and Tribal Organizations. These Agencies plan, develop and support comprehensive in-home and community services that create opportunities for active older persons and meet the needs of older persons at risk of losing their independence." The Agency organizes information on aging according to the needs of those seeking information (1) older persons and their families, (2) practitioners and other professionals, and (3) researchers and students.

- **American Association of Retired Persons (AARP)**
  http://www.aarp.org
  "AARP is a nonprofit, nonpartisan organization dedicated to helping older Americans achieve lives of independence, dignity and purpose." AARP's web page offers position statements on issues age discrimination in employment, elder abuse, generic drugs, and transportation. There are also answers to frequently asked questions such as "Is it true that AARP's directors and managers receive huge salaries and expense accounts?" and "If AARP does not endorse political candidates or parties, how do you make your presence felt before Congress and the state legislatures?"

- **Andrus Geronotolgy Library**
  http://www-lib.usc.edu/Info/Gero/geroul.htm
  The Andrus Gerontology Center Library at the University of Southern California provides links to the internet sources on aging and related topics. Of particular interest is the "Sociology and Psychology" link which covers topics such as Alzheimer's Disease and Related Disorders, Caregiving, Death & Dying,

Demographics and Statistics, Economics and Finance, Housing, Intergenerational Relations, Legal Rights, Psychology and Psychiatry, Retirement, and Social Gerontology.

- **Closing the Gap: Aging**
  http://www.omhrc.gov/ctg/ctg-agng.htm
  "*Closing the Gap* is a bimonthly newsletter from the Office of Minority Health. Each issue is devoted to a specific health topic of concern to minority communities." The May/June 1996 issue contains 20 articles related to aging. Some titles include "Minority Health Perspective: Mental Health and Older Adults," "Educating Hispanic/Latino Elderly," "American Indian/Alaska Native Elders Face Challenge of Long Term Care," and" Smoking Patterns of Older Americans."

AIDS and HIV
- **Basic Statistics on HIV and AIDS**
  http://www.cdc.gov/nchstp/hiv_aids/stats.htm
  The Centers for Disease Control posts tables containing statistics on the numbers of AIDS cases in the United States, by race or ethnicity, by exposure category, and so on. You can also find a list of the 10 states and 10 cities reporting the highest number of AIDS cases. In addition, CDC posts international projections on AIDS and HIV cases.

Americas
- **Country Health Profiles for the Americas Only**
  http://www.paho.org/english/country.htm
  (see Pan American Health Organization)

Amnesty International
- **Amnesty International Home Page**
  http://www.oneworld.org/amnesty/index.html
  Amnesty International is devoted to the cause of human rights. The group focuses on prisoners of conscience, abuse by opposition groups, asylum seekers, and those in exile.

Applied Sociology
- **A Checklist for Job Hunting and Launching a Career in Applied Sociology**
  http://www.indiana.edu/~appsoc/job.htm
  Catherine Mobley at the University of Maryland, College Park, prepared this checklist for those considering a career as an applied sociologist. She considers the advantages and purpose of five career-finding strategies: (1) networking, (2) informational gathering, (3) mentoring, (4) internships, and (5)

volunteering/community service. For each of the five strategies she answers three questions: (1) What is it? (2) Why do it? and (3) How to do it?

- **Areas of Specialization Within Sociology**
  http://www.ucm.es/OTROS/isa/rc.htm
  The International Sociological Association lists 50 different areas within sociology for which the association has established research committees. The areas include community research, future research, sociology of sport, deviance and social control, housing and built environments, sociology of disasters, international tourism, and clinical sociology. Brief descriptions of each subfield and corresponding committee objectives are posted.

- **35 Things To Think About If You Are Considering Sociology (as a Career)**
  http://www.indiana.edu/~appsoc/35things.htm
  Stephen F. Steele at Anne Arundel Community College identifies "some 'unofficial encouraging' thoughts for people planning a career in sociology." Although there are few ads in the classifieds that say, "Wanted: Sociologists," a sociology degree can help prepare you for a wide variety of jobs. However, you must take steps to articulate the value of a sociology degree and to develop practical skills.

Area Codes

- **Search for an Area Code**
  gopher://coral.bucknell.edu:4320/7areacode
  Do you need to know an area code? Submit the name of the city and state, and the 3-digit area code appears.

Best of the World Wide Web

- **Best of the Best of the Web**
  http://198.105.232.4/powered/bestofbest.htm
  This page showcases some of the most innovative sites on the Internet, but it is best viewed with a graphics-capable Internet browser.

- **Cool Site Winners**
  http://cool.infi.net/
  This page changes daily and highlights the favorite sites of celebrities. Each day a new celebrity is the "agent of cool," and a short biography of him or her is posted along with links to that celebrity's favorite sites. This is a good site for those interested in the sociology of popular culture.

- **New Sites on the Web**
  http://webcrawler.com/select/nunu.new.html
  This is a list of new sites posted on the web in the past week. The new sites are grouped under the headings "Arts and Entertainment," "Business," "Computers," "Education," "Politics," "Medicine," "Diseases," "Humanities," "Culture," "Hobbies," "Technology," "Sports," and "Travel."

- **Spider's Pick**
  http://www-scf.usc.edu/~vandenhe/spidersweb/
  The Spider's Pick of the Day is a site on the Web the "Spider" finds interesting (note: the Spider's interests are wide-ranging). In addition to the site of the day, you can view "The Spider's Past Picks List," "The Spider's Picks for the Past Month," "The Spider's Past Picks Index," and "The Spider's Random Past Picks."

- **Top Web Sites**
  http://point.lycos.com/categories/index.html
  This page gives access to the top Web sites related to a variety of general subjects such as education, government, and the world.

Bibliographic Formats

- **A Guide for Writing Research Papers Based on Modern Language Association Documentation**
  http://webster.commnet.edu/mla.htm
  This guide covers the various stages of the research process with advice on how to gather material, keep track of sources, take notes, document and cite references, and so on. There is also an informative article on plagiarism and examples of how to cite print and electronic references.

- **Bibliographic Formats for Citing Electronic Information**
  http://www.uvm.edu/~ncrane/estyles/
  Do you have questions about how to cite information from the World Wide Web? This site covers the MLA and APA citation formats for electronic information. It presents clear examples for almost any type of electronic resources. APA stands for American Psychological Association and MLA stands for Modern Language Association, and both formats are commonly used across many academic disciplines.

- **Formatting in Sociology**
  http://owl.english.purdue.edu/Files/60.html
  The Purdue On-Line Writing Laboratory posts the bibliographical format as recommended in the American Sociological Review (a journal of the American Sociological Association). It presents

clear examples of footnotes, in-text citations, page references, and almost every kind of reference —journals, books, public documents, unpublished materials—but not electronic publications.

Books Online

- **Main On-Line Books Page**
  http://www.cs.cmu.edu/Web/books.html
  This site is an index of over 1,800 online books that you can browse by author or title. Some foreign-language materials are available. Books from this index of particular interest to sociologists are listed below. Keep in mind that the books on the Internet are there because someone has taken the time to type and post them for others to use. Consequently this list represents only a very small proportion of books important to the discipline of sociology.

- **Charles Darwin**
  The Origin of the Species
  http://www.literature.org/Works/Charles-Darwin/origin/

- **Friedrich Engels**
  The Housing Question
  http://csf.colorado.edu/psn/marx/Archive/1872-HQ/
  Ludwig Feuerbach and the End of Classical German Philosophy
  http://csf.colorado.edu/psn/marx/Archive/1886-ECGP/
  The Peasant War in Germany
  http://csf.Colorado.EDU/psn/marx/Archive/1850-PWG/

- **Gustave LeBon**
  The Crowd: A Study of the Popular Mind
  http://etext.lib.virginia.edu/cgibin/toccer?id=BonCrow&tag=
      public&images=images/modeng&data=/1v1/Archive/eng-
  parsed&part=0

- **Thomas Malthus**
  An Essay on the Principle of Population
  http://socserv2.socsci.mcmaster.ca/~econ/ugcm/3ll3/malthus/
      popu.txt

- **Karl Marx**
  The Class Struggles in France, 1848 to 1850
  http://csf.Colorado.EDU/psn/marx/Archive/1850-CSF/
  The Communist Manifesto
  gopher://wiretap.spies.com/00/Library/Classic/manifesto.txt
  Wage-Labor and Capital
  http://csf.Colorado.EDU/psn/marx/Archive/1849-WLC/

- **Alexis de Tocqueville**
  Democracy in America
  http://xroads.virginia.edu/~HYPER/DETOC/home.html

- **Thorstein Veblen**
  The Theory of the Leisure Class
  http://socserv2.socsci.mcmaster.ca/~econ/ugcm/3ll3/veblen/leisure
  index.html

Calculators

- **Calculators On-Line**
  http://www-sci.lib.uci.edu/HSG/RefCalculators.html
  This site is a list of calculators for almost everything you can
  imagine. You can figure out the maximum hull speed on your
  sailboat, determine your financial net worth, calculate your body
  mass, do simple and complex math, and much more.

Calendars

- **Calendar Generator**
  http://www.stud.ntnu.no/USERBIN/steffent/kalender.pl
  Are you interested in knowing what day your birthday will be on
  in the year 2005? Or do you need to know what day of the week it
  was when Abraham Lincoln was born on February 12, 1809? You
  can submit a year between 1754 and 3000 to generate a 12-month
  calendar for that year.

Career Guides and Opportunities

- **Career Connections**
  http://www.career.com/
  From this site it is possible to learn about job openings from
  around the country and get in touch with employers. You can
  search for jobs by company, category, or location, and this site
  profiles jobs for new graduates. There is a link to "Hot Jobs,"
  which lists immediate openings and the qualifications needed for
  them.

- **Career Development Manual**
  http://www.adm.uwaterloo.ca/infocecs/CRC/manual-home.html
  The University of Waterloo posts the *Career Development
  Manual*, which covers the "Steps to the Right Job": (1) self-
  assessment, (2) researching occupations, (3) decision making
  about long- and short-term occupational goals, (4) taking the steps
  to find employment, and (5) evaluating job offers and accepting
  the job.

- **Career Magazine**
  http://www.careermag.com/
  > This site provides access to over 14,000 current job listings from around the world. It also allows you to post your résumé online, and provides information on companies that recruit college graduates, and informative news articles on career opportunities.

- **CareerPath**
  http://192.216.189.91/res/owa/rb_applicant.display_rblogin?
  > From this site it's possible to search newspaper employment ads from 10 major cities. There is no fee charged for this service, but you must register to use it.

- **Career Shop's Résumé, Job and Employment Site**
  http://www.careershop.com/
  > This site is "an on-line database of résumé profiles and employment opportunities, designed to assist job seekers and hiring employers alike. Job seekers can post résumé profiles in our résumé database and perform job searches of our Job Openings database—online and free of charge!" There is also advice on preparing for an interview and promoting yourself to employers.

- **ERIC Digests**
  > ERIC is a clearinghouse for educational material. The following are selected ERIC publications related to jobs/careers. The URL for ERIC is gopher://INET.ed.gov:12002 . Once you are in the site, select "ERIC.src." Once in the ERIC Web site, you will be prompted to enter a keyword. Use the numbers to the left of the titles listed below as the keyword for immediate access. Occasionally the number will not work. When this happens, enter the title at the query prompt.

  > **ED292974** Workplace Literacy Programs
  > **ED346318** Job Search Methods
  > **ED376274** Job Search Skills for the Current Economy

- **Occupational Outlook Handbook**
  http://stats.bls.gov/ocohome.htm
  > The Bureau of Labor Statistics publishes this career guidance handbook that lists comprehensive job descriptions for hundreds of occupations, including their educational and other qualifications, working conditions, earnings, employment trends, and job prospects. Users of *The Occupational Outlook Handbook* can browse by job title or search by keyword, such as "social scientist." The handbook is revised every two years.

Careers in Sociology

- **A Checklist for Job Hunting and Launching a Career in Applied Sociology**

  http://www.indiana.edu/~appsoc/job.htm

  Catherine Mobley at the University of Maryland, College Park, prepared this checklist for those considering a career as an applied sociologist. She considers the advantages and purpose of five career-finding strategies: (1) networking, (2) informational gathering, (3) mentoring, (4) internships, and (5) volunteering/community service. For each of the five strategies she answers three questions: (1) What is it? (2) Why do it? and (3) How to do it?

- **Areas of Specialization Within Sociology**

  http://www.ucm.es/OTROS/isa/rc.htm

  The International Sociological Association lists 50 different areas within sociology for which the association has established research committees. The areas include community research, future research, sociology of sport, deviance and social control, housing and built environments, sociology of disasters, international tourism, and clinical sociology. Brief descriptions of each subfield and corresponding committee objectives are posted.

- **35 Things to Think About If You are Considering Sociology (as a Career)**

  http://www.indiana.edu/~appsoc/35things.htm

  Stephen F. Steele at Anne Arundel Community College identifies "some 'unofficial encouraging' thoughts for people planning a career in sociology." Although there are few ads in the classifieds that say, "Wanted: Sociologists," a sociology degree can help prepare you for a wide variety of jobs. However, you must take steps to articulate the value of a sociology degree and to develop practical skills.

Career Placement

- **Jobs**

  http://www.agen.tamu.edu/links/jobs.html

  This site contains a listing of universities and corporations with available jobs, and it tells you how to contact employment services, employment recruiters, and the like.

- **JobTrack**

  http://www.jobtrak.com/

  The creators of this site call it "the premier site for recruiting college students and recent graduates. The company has formed partnerships with over 300 college and university career centers and is utilized by over 150,000 employers." There is a section on graduate schools that includes advice on how to write the essays that are part of graduate school applications and information on obtaining grants to help pay for graduate study.

- **Major Resource Kit**
  http://www.udel.edu/CSC/
  The University of Delaware has compiled a "Major Resource Kit" for 49 undergraduate programs to help new graduates in their search for a job. Each kit includes information such as the job titles recent graduates have attained that directly relate to their major. This site tells the amount of education required for specific jobs, and it also gives tips for increasing employability. A few words of warning: the kinds of jobs listed may not be representative of the jobs for which a major may be eligible to apply. For example, the jobs listed for sociology clearly understate the wide range of occupations sociologists hold.

Carter Center

- **Carter Center**
  http://www1.cc.emory.edu/CARTER_CENTER/
  The Carter Center, founded by former President Jimmy Carter, is dedicated to fighting disease, hunger, poverty, conflict, and oppression by working for development, urban revitalization, and global health.

Census Data

- **Census Bureau Press Releases**
  http://www.census.gov/Press-Release/www
  The press releases at this site can be searched by subject or by date. Many of the press releases announce new statistics related to a wide range of subjects, from aging to national population estimates.

- **Census 2000 Press Releases**
  http://www.census.gov/Press-Release/www/2000.html
  This Web site contains Census Bureau press releases relating to Census 2000. Press releases announce plans for making the census simpler, less expensive, and more accurate and summarize recommendations made by various advisory committees. Most of the press releases relate to issues of racial and ethnic classification and to plans to reach populations that have traditionally been undercounted.

- **Demographic and Population Contacts at the Census Bureau**
  http://www.census.gov/contacts/www/c-demopop.html
  Do you need information from the Census Bureau but are unsure of where to look for it? This site lists the names and phone

numbers of people by department at the Census Bureau who can help you find information on topics ranging from the age structure of the U.S. population to voter characteristics.

- **Search the Census Bureau**
  http://www.census.gov/main/www/srchtool.html
  This Census Bureau Web site allows you to search its ever-expanding library of online documents. Just plug in a keyword and "matching" documents from which to select are listed.

- **United States Bureau of the Census**
  http://www.census.gov/
  The Bureau of the Census is a general-purpose federal agency that collects, tabulates, and publishes a wide variety of statistical data about the people and economy of the United States. The statistical results of the censuses, surveys, and other Census Bureau programs are available to the public.

Children

- **Administration for Children and Families Press Releases**
  http://www.acf.dhhs.gov/news/
  This site is maintained by the Department of Health and Human Services. These press releases offer statistics on children and information about demonstrations related to welfare reform.

- **State of the World's Children**
  http://www.unicef.org/sowc96/contents.htm
  This site provides links to resources such as statistical tables for measuring children's well-being and statistics on children from regions around the world. There is also information about children affected by war and a report on the improvements that have taken place in the 50 years that UNICEF has been involved with children's needs.

City-Level Information (U.S. and World Cities)

- **City Data**
  http://www.lib.virginia.edu/socsci/ccdb/city94.html
  The University of Virginia Social Sciences Data Center provides access to the 1994 County and City Data Book. At this URL city-level information on the foreign-born population, family and nonfamily households, number of vehicles per household, and so on, is available. There are approximately 240 data tables from which to select.

- **Top City Rankings**
  http://www.census.gov/stat_abstract/ccdb.html

This Census Bureau site contains tables in which U.S. cities are ranked according to factors such as percentage of foreign-born residents, population size, percentage of workers using public transportation, and so on.

- **USA CityLink**
  http://banzai.neosoft.com/citylink/
    The USA CityLink is a collaborative project that claims to offer the most comprehensive list of links to city and state Web sites. Many U.S. cities participate in this project. CityLink requires that 90 percent of the information each city and state links to the host site must be about the state or city, and it must be useful to someone planning to move to, travel to, or visit the area. Examples of useful information include school system profiles, historical overviews, and descriptions of interesting tourist attractions and cultural events.

Cocaine and Federal Sentencing Policy

- **Special Report to Congress: Cocaine and Federal Sentencing Policy**
  http://www.acsp.uic.edu/lib/ussc/chapter1.htm
    This site contains the *Special Report to the Congress: Cocaine and Federal Sentencing Policy (as Directed by Section 280006 of Public Law 103-322)*. This report seeks to determine if federal sentencing guidelines and policies for cocaine offenses are fair and effective. Currently mandatory minimum prison sentences are based on the specific quantity of drug being distributed. Under current policies crack cocaine is treated differently from powder cocaine such that possession of 500 grams of powder triggers a five-year minimum mandatory sentence while possession of only 5 grams of crack triggers the same sentence. The report considers whether crack cocaine is different enough from powder cocaine to warrant the harsher penalty. This report contains eight chapters and three appendices, which can be accessed from this URL.

Code of Ethics (American Sociological Association)

- **Code of Ethics of the American Sociological Association**
  http://www.asanet.org/ethics.htm
    The Code of Ethics is a set of rules and general principles that guide sociologists in the areas of research, teaching, and practice. The code " attempts to meet the expressed needs of sociologists who have asked for guidance in how best to proceed in a variety of situations involving relations with respondents, students, colleagues, employers, clients, and public authorities."

Colleges and Universities

- **Academe This Week**
  http://chronicle.merit.edu/.almanac/.links.html
  This is a statistical portrait of higher education in the United States. It includes statistics on graduation rates at NCAA Division I institutions, the number of college degrees awarded, the average pay of full-time professors, and enrollment by age, sex, and race.

- **List of American Universities Home Pages**
  http://www.clas.ufl.edu/CLAS/american-universities.html
  This site gives the homepages of American universities granting bachelor or advanced degrees. It provides links to international universities, Canadian universities, and community colleges as well.

- **Peterson's Education Center**
  http://www.petersons.com/
  (see Education)

College Experience

- **Campus Newspapers on the Internet**
  http://www.lib.virginia.edu/cataloging/vnp/college.html
  This site is posted by *The Daily Beacon,* the student newspaper at the University of Tennessee, Knoxville. In addition there are links to hundreds of other college newspapers. Depending on the site, you can access the current issue, recent issues, and/or archived issues.

- **Preparing Your Student for College**
  http://www.ed.gov/pubs/Prepare/
  This U.S. Department of Education guidebook covers areas such as choosing a college, financing an education, and doing long-range planning. The guidebook also lists and defines important terms related to the college experience, such as B.A. and B.S.

- **ERIC Digests**
  ERIC is a clearinghouse for educational material. The following are selected ERIC publications related to study skills, and the college experience. The URL for ERIC is gopher://INET.ed.gov:12002. Once you are in the site select "ERIC.src." Once in the ERIC Web site, you will be prompted to enter a keyword. Use the numbers to the left of the titles listed below as the keyword for immediate access. Occasionally the number will not work. When this happens, enter the title at the query prompt.
Study Skills
  ED250694 Qualities of Effective Writing Programs

ED250696 Vocabulary

ED291205 Critical Presentation Skills—Research to Practice

ED296347 Audience Awareness: When and How Does It Develop?

ED300805 Note-Taking: What Do We Know About the Benefits?

ED301143 Learning Styles

ED302558 Improving Your Test-Taking Skills
Critical Thinking

ED318039 How to "Read" Television: Teaching Students to View TV Critically

ED327216 Information Skills for an Information Society: A Review

ED326304 How Can We Teach Critical Thinking?

ED372756 Information Literacy in an Information Society

ED385613 Making the A: How to Study for Tests
College

ED284510 Self-Study in Higher Education: The Path to Excellence

ED266339 Selecting a College: A Checklist Approach

ED284514 Student Stress: Effects and Solutions

ED284526 Reducing Stress Among Students

ED286938 Alternatives to Standardized Tests

ED351079 First-Generation College Students

Congressional Activity

- **Almanac of American Politics**
  http://www.archive.org/pres96/websites-mar/politicsusa/almanac.html

  The *National Journal* maintains this site, which lists U.S. representatives and senators for each state along with some background information on each person. It also gives information on the senators and representatives who rank highest in fundraising and spending, who receive the most PAC and individual contributions, and who have the most cash on hand.

- **Cover Stories from Previous Congressional Quarterly Weekly Report Stories**
  http://www.cq.com/wr.htm

  From this site it's possible to read the cover stories from the *Congressional Quarterly Weekly* for any week in the previous year. The articles deal with political issues, most of which are in the news.

- **THOMAS: Legislative Information on the Internet**
  http://thomas.loc.gov/

  This site is an attempt to make federal legislative information freely available to the Internet public. "Hot Topics" are those bills and amendments that are the subjects of floor action, debate, and hearings in Congress and that are frequently reported on by the popular media. You can search through topics such as foreign aid and urban affairs for relevant current legislation. From this site it is possible to read the full text of bills and find out who sponsored and cosponsored them.

Consumer Information

• **Consumer Information Center**
  http://www.pueblo.gsa.gov/textonly.htm
  This page is maintained by the Consumer Information Center of the U.S. General Services Administration. It allows you to read reports on consumer goods such as cars and toys.

Copyright

• **Copyright Website**
  http://www.benedict.com
  Benedict O'Mahoney, corporate counsel for a high-tech software company, created this Web site for anyone in need of practical and relevant copyright information. Among other things, he covers the basics of copyright law, famous copyright infringement cases, copyright registration, and copyrighted materials on the Web. O'Mahoney welcomes dialogue and suggestions that address the "myriad of copyright tangles that currently permeate the Web."

Country Background Notes

• **Background Notes on the Countries of the World**
  http://www.state.gov/www/background_notes/
  (see Country-Level Information)

Country Codes for Internet URLs

• **ISO Country Codes:**
  http://www.nw.com/zone/iso-country-codes
  Use this site if you want to interpret the country code on an Internet URL address. An example is the National Library of Australia (http://www.nla.gov.**au**/gov/govinfo.html). The **au** in this index tells you that the site is from Australia.

Country-Level Information

• **Background Notes on the Countries of the World**
  http://www.state.gov/www/background_notes/index.html
  This site contains statistical and general information on most of the countries of the world (but not the United States) and covers geography, people, education, economics, and membership in international organizations.

- **Contemporary Conflict**
  http://www.cfcsc.dnd.ca/links/wars/index.html
    Contemporary Conflict is a compilation of links related to countries experiencing some type of conflict. The amount and quality of data available varies by country, but often these links lead to information on the history of the conflict, to documents and articles important in understanding the conflict, and to relevant organizations and agencies.

- **Country Studies**
  http://lcweb2.loc.gov/frd/country.html
    This site gives access to book-length information on Ethiopia, China, Egypt, Indonesia, Israel, Japan,the Philippines, Singapore, Somalia, South Korea, and Yugoslavia. The site eventually will include over 60 country studies written by a multidisciplinary team of authors. It is a comprehensive source of information about all areas of life including politics, economics, culture, religion, population, history, and culture.

- **Country Reports on Human Rights Practices**
  http://homer.louisville.edu/groups/library-www/ekstrom/govpubs/
  goodsources/hrightsreports.html
    (see Human Rights)

- **International Demographic Data**
  http://www.census.gov/ftp/pub/ipc/www/idbsum.html
    (see Population, Countries)

- **National Library of Australia Internet Site**
  http://www.nla.gov.au/gov/govinfo.html
    This site contains information about and from the governments of Australia, New Zealand, and selected countries, territories, and states within Asia, Canada, the United Kingdom, the United States, Europe, Africa, and the Middle East. The information available on each country varies.

- **1997 CIA World Factbook**
  http://www.odci.gov/cia/publications/factbook/
    The *World Factbook* contains a wide range of information on all countries and bodies of water. There are brief summaries and statistics related to topics such as the unemployment rate, population size, and total fertility.

- **Population for the Countries of the World**
  gopher://gopher.undp.org:70/00/ungophers/popin/wdtrends/
  pop1996.txt
    (see Population, Countries)

- **Press Releases by Country**
  http://www2.iadb.org/prensa/PCOUNTRY.HTM
  This site is a compilation of press releases from many countries. This is not a comprehensive list of countries or press releases, however, and some of the press releases are several months old.

- **Travel Warnings and Consular Information Sheets**
  http://www.stolaf.edu/network/travel-advisories.html
  (see Travel)

- **World Abortion Policies**
  gopher://gopher.undp.org:70/00/ungophers/popin/wdtrends/ charts
  The United Nations has gathered information showing the grounds on which abortion is permitted in almost every country in the world. Those grounds considered are " to save the mother's life, to preserve the mother's physical health or mental health, rape and incest, fetal impairment, economic reasons, and on request."

- **World Constitutions**
  gopher://wiretap.spies.com/11/Gov/World
  This site provides links to many of the world's constitutions and other important historical documents in their entirety.

County-Level Information

- **County Business Patterns**
  http://www.census.gov/Press-Release/www/cbp.html
  The Census Bureau gives employment numbers for states and counties according to economic sectors (agriculture, mining, construction, manufacturing, transportation, public utility, wholesale trade, finance, insurance, real estate, and services). The report also includes the number of establishments for each economic division and the total annual payroll. Read " General Explanation of CBP Series" for an explanation of how to interpret tables. Information is presented according to geographical region, state, and county. Choose " 1992-1991 State Data by 2-Digit SIC, and by County" to get county data.

- **County Demographic Profiles**
  http://govinfo.kerr.orst.edu/usaco-stateis.html
  This site is part of the U.S. Government Information Sharing Project. It allows you to create a " summary report" for any county in the United States. The summary report includes data on the total resident population (1992), percentage of the population under age 18, percentage of owner occupied housing units, median income,

unemployment rate, and elementary, high school, and college enrollment figures.

- **1994 County Files**
  http://www.lib.virginia.edu/socsci/ccdb/county94.html
  The University of Virginia Social Sciences Data Center provides access to the 1994 *County and City Data Book.* Information such as the foreign-born population, family and nonfamily households, number of votes cast for president in 1992, and so on, is available for every county in each of the 50 states. There are approximately 240 data tables from which to select.

- **See If Your County Lost or Gained Population 1990–1995**
  http://www.usatoday.com/news/states/special/pop/npop000.htm
  (see Population, County)

Crime

- **Department of Justice**
  http://www.usdoj.gov/
  The DOJ is responsible for protecting American citizens from criminals and subversive activities, for ensuring fair competition among businesses, for safeguarding the consumer, and for enforcing drug, immigration, and naturalization laws. The DOJ represents the U.S. government in legal matters such as Supreme Court lawsuits in which the government is involved.

- **World Crime Survey Data**
  http://www.ifs.univie.ac.at/~uncjin/wcs.html
  Data sets from the United Nation's *World Crime Survey of Crime Trends and Criminal Justice* are available at this site. Scroll down until you see " The Fourth United Nations Survey of Crime Trends and Operations of Criminal Justice Systems: Available Online Data Set" to find the codebook, which describes the research design and defines key terms. There are more than a hundred tables related to 13 subjects including police personnel, prisoners and prison sentences, prosecutions by age and sex, and average sentence length. The UN survey was distributed to all 134 member states, and 98 responded. Because responding governments did not supply data for every question, the tables vary in the number of countries covered.

Culture

- **Cultural Exchange**
  http://deil.lang.uiuc.edu/exchange/
  Exchange is a University of Illinois, Urbana-Champaign, electronic publication that welcomes submissions of 2,000 words or less from non-native English speakers from around the world.

This publication includes essays in which authors describe some aspect of their culture (a graduation ceremony, a ritual, a holiday celebration, a tradition), react to news events in their country, submit a short story, or present a poem. Although the primary goal of *Exchange* is to support a forum for non-native English speakers to express themselves in English, the journal is also a unique resource for English speakers interested in culture.

Currency Converter

- **Currency Converter**
  http://www.xe.net/currency/
  Select the country that uses the currency you wish to have converted, and the value of all other countries' currencies will be calculated in relation to that currency.

Defense

- **Department of Defense**
  http://www.dtic.dla.mil/defenselink
  The Department of Defense (DOD) is in charge of securing and training the military personnel needed to prevent war and protect the security of the country and its interests. The major divisions of the U.S. military are the Army, Navy, Marine Corps, and Air Force.

- **NATO Press Releases**
  http://www.nato.int/docu/pr/pr96e.htm
  The North Atlantic Treaty Organization (NATO) is an alliance for collective defense formed in 1949 and linking 14 European countries, the United States, and Canada. Since 1989 (the fall of the Berlin Wall and the symbolic fall of communism) NATO has worked to establish cooperation between the governments of Central and Eastern European countries and with the newly independent states of the former Soviet Union. Its press releases focus on activities related to this task.

Demographic Profile Of The United States

- **American Demographics Magazine**
  http://www.demographics.com/Publications/AD/Index.HTM
  American Demographics Magazine presents articles at this site dealing with demographic and marketing issues such as aging, racial classification, home workers, illegal immigration, consumer problems, and markets. You have the option of searching through each issue or of doing a keyword search.

- **Population Profile of the U.S.: 1995**
  http://www.census.gov/ftp/pub/prod/www/abs/mspop04a.html

This Census Bureau Web site contains a series of selected facts about the U.S. population. The site lists statistical facts (such as the percentage of 3- and 4-year-olds enrolled in nursery school, the share of households occupied by families, and so on). Individually the facts might seem trivial, yet collectively they offer an interesting profile of the U.S. population. Additional selected statistics on topics such as the percentage of the population with a high school diploma and the percentage of people living in urban areas are available at gopher://gopher.census.gov/00/Bureau/Pr/Subject/Pop/cb94-34.txt

- **Population Reference Bureau Releases**
  http://www.prb.org/media/pressrel.htm
  The Population Reference Bureau's press releases focus on population-related issues (births, deaths, migrations) that affect life in the United States.

- **Statistical Abstract Frequently Requested Population Tables**
  http://www.census.gov/stat_abstract/pop.html
  This Census Bureau site gives population totals for the resident U.S. population (1900–1994), as well as population counts for states, large metropolitan areas, and cities with 100,000 or more people. The population totals are subdivided according to sex, age, race, and household type.

Dictionaries Online

- **Easton's Bible Dictionary**
  http://ccel.wheaton.edu/easton/ebd/ebd.html
  This site is helpful when you come across a biblical reference but are unsure of the meaning or when you are curious about how the Bible treats a subject such as adoption or war.

- **Webster's English Dictionary**
  http://work.ucsd.edu:5141/cgi-bin/http_webster
  Simply enter the word for which you need a definition and select "Lookup definition."

Disabled

- **The Disability Rag**
  http://www.etext.org/Politics/Disability.Rag/
  This site gives access to *The Disability Rag*, an online magazine that deals with issues disabled people face, such as access to public accommodations and discrimination in hiring practices.

Economic Indicators

- **Economic Statistics Briefing Room**
  http://www.whitehouse.gov/fsbr/esbr.html
  This page provides access to current federal economic indicators such as disposable personal income, civilian labor force, and consumer price index. It also provides links to information produced by a number of federal agencies such as the Bureau of Economic Analysis, U.S. Census Bureau, Federal Reserve Board, and Bureau of Transportation Statistics.

Economic Policy

- **Country Reports on Economic Policy and Trade**
  http://www.state.gov/www/issues/economic/index.html
  (see Trade Practices)

Education

- **College Enrollment and Work Activity of 1995 High School Graduates**
  http://www.bls.census.gov/cps/pub/hsgec_1095.htm
  The Bureau of Labor Statistics posts information on the number of students who dropped out of high school between October 1994 and October 1995, as well as the number who graduated in 1995. There is also information on the labor force status of 1995 high school graduates, 1994–95 college dropouts, and persons ages 16–24.

- **Condition of Education**
  http://www.ed.gov/pubs/CondOfEd_95/
  The Condition of Education is a 1995 report prepared by the U.S. Department of Education National Center for Educational Statistics. The full text of the report is online, and it covers these kinds of topics: " Are More Young People Going to College?" " The Cost of Higher Education," and " What Do We Know About the Quality of Schools?"

- **Department of Education**
  http://www.ed.gov/
  The mission of the Department of Education is to ensure equal access to education and promote educational excellence in schools throughout the United States. It establishes policy and administers and coordinates most federal assistance to education.

- **Education Attainment—Historical Tables**
  http://www.census.gov/population/socdemo/education/
  This Census Bureau site presents data on " Educational Attainment of Persons 25 Years and Over By State: 1990, 1980, 1970," " Percent of Persons 25 and Over Who Have Completed High

School or College: Selected Years 1940 to 1993," and "Mean Earnings of Workers 18 Years Old and Over, by Educational Attainment: 1975 to 1992."

- **Education Policy Analysis Archives**
  http://olam.ed.asu.edu/epaa/
  This site contains an online scholarly journal focusing on education policy at all levels and in all countries.

- **Peterson's Education Center**
  http://www.petersons.com/
  Peterson's is one of the leading publishers of information on U.S. educational opportunities at all levels. Links are grouped according to the following headings: "K-12 Schools," "Colleges & Universities," "Graduate Study," "Studying Abroad," "Careers & Jobs," "Summer Programs," "Language Study," "Executive Education" (short-term programs targeted at top management), "Distance Learning" (electronically delivered courses), "Continuing Education," "Learning Holidays" (camps, such a golf and tennis), "Testing & Assessment," "Vocational-Technical Schools," and "Metro Guides" (educational opportunities, such as daycare and adult daycare, in large metropolitan areas). While this Web site is designed to create interest in Peterson Publications by giving brief overviews of the kinds of information that can be found in publications that are for sale, the site offers excellent general descriptions of a range of educational opportunities.

- **School Enrollment—Historical Tables**
  http://www.census.gov/population/www/socdemo/sch94.html
  This site contains seven tables related to nursery school through university enrollment and dropout rates since 1947 for various age groups and for males and females.

- **U.S. Department of Education Publications**
  http://www.ed.gov/publications.html
  This site is provided by the U.S. Department of Education and offers access to many of its publications, including newsletters, publications for parents, education statistics, and so on.

- **Worldspeaker**
  http://www.tiac.net/users/worldspe/
  Caput Mundi Incorporated posts this site which serves as a forum for international educational institutions to share ideas and news of their activities in order to promote awareness of global educational developments.

Elderly

- **Data on the Elderly**
  http://aspe.os.dhhs.gov/96gb/aappend.txt
  This compilation of statistics from government agencies such as the Bureau of Labor Statistics and the U.S. Bureau of the Census "presents historical and current data on the demographic and economic characteristics of the elderly, including information on population, life expectancy, labor force participation, marital status, living arrangements, poverty rates, and income." Data related to health care for the elderly can be found at http://aspe.os.dhhs.gov/GB/apenb.txt

- **Health Status of the Elderly**
  http://aspe.os.dhhs.gov/96gb/bappend.txt
  This Department of Health and Human Services site reports on the health status of the elderly (including self-perceived health status), health insurance coverage, and expenditures. Many of the tables are so wide that they wrap around to the next line, making them difficult to read unless you have Netscape. However, the information in the tables is summarized in words.

E-mail Addresses

- **Finding an E-mail Address**
  http://sunsite.oit.unc.edu/~masha/
  Do you need to find someone's e-mail address or check to see if they have one? This University of North Carolina site uses a question-and-answer format to help you find an address. Examples of questions that help to narrow your search are "What region or country is he/she located in?" and "Is he/she on a network other than the Internet?"

Encyclopedias Online

- **Encyclopedia Smithsonian**
  http://www.si.edu/resource/faq/start.htm
  Topics included in this encyclopedia are determined by public demand for information on a topic. As of June 1996, there were nine general topics: armed forces history, anthropology, mineral sciences, musical history, physical sciences, conservation of textiles, transportation history, and vertebrate zoology.

- **Free Internet Encyclopedia**
  http://clever.net/cam/encyclopedia.html
  Creators of this site suggest that a more accurate name than "Free Internet Encyclopedia" is "Free Internet Encyclopedia Index" because this is an encyclopedia of information on the Internet. This encyclopedia has two divisions: Macroreference (large general topics such as Africa, courts, and so on) and Microreference (short

bits of information on specific topics such as asthma, Jane Austen, and so on).

English as a Second Language

- **English as a Second Language Home Page**
  http://www.lang.uiuc.edu/r-li5/esl/
  Whether you are someone who is learning English as a second language, a native speaker who needs to brush up on vocabulary, or someone who is interested in how the English language is presented to foreign-language speakers, this site is beneficial. This site presents links related to "Listening and Speaking," "Reading," and "Writing." Your computer must have audio capabilities in order for you to use "Listening and Speaking."

Environment

- **Disaster Reports**
  http://www.ifrc.org/pubs/wdr/96/
  The U.S. Agency for International Development posts fact sheets describing the effects of disasters such as hurricanes, floods, civil wars, cyclones, and fires occurring in countries around the world.

- **World Resources 1996-1997**
  http://www.wri.org/wri/wr-96-97/index.html
  This World Resources Institute Web site offers access to hundreds of country- and regional-level studies of the condition of the environment and the state of natural resources. Reports are available on the following regions and countries within each region: Africa, Asia, the Caribbean, Central America, Eastern Europe, Europe, new independent states, North America, Oceania, South America, Commonwealth States, Latin America, and the Mediterranean.

- **Environmental Protection Agency**
  http://www.epa.gov/
  The EPA's mission is to regulate and enforce laws regarding the use and disposal of solid wastes, pesticides, radioactive materials, and toxic substances that pollute the air and water.

- **Environmental Protection Agency (EPA) Journal**
  http://www.epa.gov/docs/epajrnal/
  The *EPA Journal* is published quarterly. The articles in each issue focus on an environmental theme such as Earth Day or environmental awareness.

- **Global Stewardship Network**
  http://www.iisd.ca/updates/

This is a free news service covering issues related to global stewardship (that is, "caring for the earth and its current inhabitants, as well as a responsibility to leave to future generations a planet capable of sustaining life"). Past issues are also available at this site.

- **Greenpeace**
  http://www.greenpeace.org/campaigns2.html
  Greenpeace is an independent activist organization that confronts global environmental problems through nonviolent action, with the goal of forcing action to protect the earth's ability to nurture life in all its diversity.

- **State of the World Indicators**
  http://www.igc.apc.org/millennium/inds/
  This site provides links to various indicators for a quick overview of the state of the world's environment. Some indicators are water availability, species extinctions per day, and years until half of known crude oil is gone. There is information about each indicator as well.

- **World Resources Institute News Release**
  http://www.wri.org/wri/press/wr96-nr.html
  "The mission of the World Resources Institute (WRI) is to move human society to live in ways that protect the Earth's environment and its capacity to provide for the needs and aspirations of current and future generations." The news releases for 1996–1997 focus on environmental imbalances associated with urbanization.

Ethnic Classification

- **Race and Ethnicity Standards for the Classification of Federal Data on Race and Ethnicity**
  http://www1.whitehouse.gov/WH/EOP/OMB/html/fedreg/race-ethnicity.html
  (see Statistical Policy Directive No. 15)

Etiquette on the Net

- **Emoticons and Abbreviations**
  http://pw2.netcom.com/~jampie96/emoticons.html
  Do you sometimes get the feeling that people misunderstand your e-mail messages? Then this is the site for you. Put an emoticon at the end of a sentence to show how you really feel. Let people know if you are on the verge of tears or if you are so happy you are "bug-eyed." The combination of keys needed to make the face to fit the message are described.

- **Internet Terms**
  http://134.84.217.16/icec/training/terms.html
  > Stop guessing about the meaning of Internet terms and acronyms!
  > This site lists and defines common abbreviations and terms from
  > " Access" to " WWW."

- **Netiquette**
  http://email.miningco.com/msub1.htm
  > Did you know that you can be rude without meaning to be? This
  > site teaches new and/or unsophisticated users "netiquette,"
  > etiquette of the 'net. This site covers the rules governing proper
  > communication, the way to convey emotion, and the most
  > productive approach to take when in need of technical support.

Experts, Authorities, and Spokespersons
- **Yearbook of Experts, Authorities, and Spokespersons**
  http://www.yearbook.com/
  > The Broadcast Interview Source posts this Web site, which offers
  > the addresses of experts, authorities, and spokespersons for various
  > organizations. Search by topic or keyword. For example, the word
  > "refugees" produced six sources of possible information, each of
  > which included an address, a profile, and a home page if
  > applicable.

FAQs
- **Usenet FAQs**
  http://www.cis.ohio-state.edu/hypertext/faq/usenet/
  > This document contains an extensive list of Usenet Frequently
  > Asked Questions (FAQs) on more than 600 subjects including
  > blacklisted Internet advertisers, body building, genealogy, junk
  > mail, medicinal herbs, the Quaker religion, table tennis, and the
  > World Wide Web. The list is alphabetized by topic. You can
  > search by newsgroup name, archive name, subject, or keyword.

Federal Courts
- **Understanding the Federal Courts**
  http://www.uscourts.gov/understanding_courts/899_toc.htm
  > Do you have questions about how the federal court system works?
  > This site provides answers to your questions with a glossary of
  > terms; an overview of the organization, operation, and
  > administration of the court system; a list of the location and
  > number of judges who sit on each court; and charts showing the
  > structure of the federal court system and the path a case takes as it
  > works its way through this system.

Financial Aid

- **Funding College**
  http://www.ed.gov/prog_info/SFA/FYE/index.html
    This U.S. Department of Education publication provides information about federal financial aid programs for college students.

- **Scholarships and Fellowships**
  http://web.studentservices.com/search/
    Searching for money to fund education is as easy as entering your name, address, and major. A list of scholarships for which you might be eligible will appear on the screen, and information about other scholarships for which you might be eligible will be e-mailed to you as they become available.

Foreign Language

- **Language**
  http://www.travlang.com/languages/
    Do you need to talk to someone who speaks a foreign language? Identify the language(s) you speak and the one that you want to learn. The computer will display common words and phrases such as " yes," " no," and " you're welcome," and words and phrases that will be useful when shopping, asking for directions, establishing a time and place to meet, and so on.

Foreign Language Dictionaries

- **Online Dictionaries**
  http://pla-net.net/resources/dict.html
    This is an extensive list of foreign-language dictionaries, from Arabic vocabulary lists to a Welsh–German dictionary.

Foreign Newspapers

- **Library of Congress Foreign Newspapers**
  http://lcweb.loc.gov/rr/news/oltitles.html#forn
    On this site the newspapers are listed alphabetically. Because this site is so large, you'll do better if you know the name of the paper you're looking for or if you have time to look through the list for a particular country. Some of the newspapers are in English.

- **Newspaper Listing—Worldwide**
  http://www.dds.nl/~kidon/papers.html
    This site provides links to numerous foreign newspapers. Most of the newspapers are in their respective languages, but some of them are in English.

Gender

- **Demographic and Health Survey**
  http://www2.macroint.com/dhs/

    The Demographic and Health Surveys (DHS) program,
    implemented by Macro International Inc. and funded primarily
    by the U.S. Agency for International Development (USAID),
    assists countries in "conducting national surveys on fertility,
    family planning, maternal and child health, and household living
    conditions" in order to obtain information on the "reproductive
    and health behavior of women throughout sub-Saharan Africa,
    the Near East, North Africa, Asia and Latin America." This Web
    site posts statistics on birth control methods, infant mortality
    rates, percentage of children immunized, and other reproductive
    health-related topics. DHS presents these statistics in the form of
    press releases. DHS also publishes a newsletter available online
    which includes articles on reproductive health. For example, the
    Vol. 7 No 1 issue includes articles such as "Fertility and Family
    Planning in Egypt" and "Safe Motherhood in the Philippines."
    This site must be viewed through Netscape.

- **Glass-Ceiling Commission Reports**
  gopher://gopher.etext.org/11/Politics/Womens.Studies

    This web site contains the U.S. Glass Ceiling Report, more
    formally known as *A Question of Equity: Women and the Glass
    Ceiling in the Federal Government*, submitted to the president of
    the United States and Congress. The report asks why there are so
    few women in top-level civil service positions (1 in 10) when they
    make up 50 percent of the white-collar employees. The report
    identifies the barriers, unrelated to women's personal career
    decisions or qualifications that play a role in this imbalance.

- **Institute for Women's Policy Research**
  http://www.iwpr.org/

    The Institute for Women's Policy Research (IWPR) is an
    independent research organization established with the goals of
    addressing the limited amount of policy-relevant research on
    women's lives and of stimulating the debate on issues of critical
    importance to women. One especially informative report on this
    website is "Status of Women in the States" which compares the
    status of women in the fifty states and the District of Columbia on
    the following areas: political participation, health, employment
    and earnings, economic autonomy, and reproductive rights. The

Institute lists titles and abstracts of its publications in the following areas: poverty and welfare, employment opportunities, work and family, health and reproductive issues, and economic issues. The IWPR Briefing Paper, "The Wage Gap: Women's and Men's Earnings" contains tables comparing women's earnings as a percentage of men's, median annual earnings by dollar over time, race and ethnic differences in earnings, and earning differences by education, occupation, and age.

- **Progress of Nations, 1996**
  http://www.unicef.org/pon96/contents.htm
    UNICEF posts the report *Progress of Nations, 1996,*which contains articles, statistics, charts, and commentary on (1) women's physical well-being, (2) nutrition/malnutrition, (3) health, with emphasis on immunizations, (4) education of women, (5) the Convention on the Rights of Children, and (6) children's well-being in the industrial world.

- **Society for the Scientific Study of Sexuality (SSSS) Web page**
  http://www.ssc.wisc.edu/ssss/
    The Society for the Scientific Study of Sexuality (SSSS), founded in 1957, is an international organization "dedicated to the advancement of knowledge about sexuality" and to "promoting human welfare by reducing ignorance and prejudice about sexuality." The SSSS believes in the "importance of both the production of quality research and the clinical, educational, and social applications of research related to all aspects of sexuality." This site contains the Statement of Ethics adopted by the SSSS, which offers a good overview of the topics covered under the subject of sexuality.

- **U.S Department of Labor: Women's Bureau**
  http://gatekeeper.dol.gov/dol/wb
    If you are interested in women's labor statistics and issues the Women's Bureau is an important resource. Particularly useful information is available under "Media Releases" and "Fact Sheets About Women in the Workplace" which contain reports on women who maintain families, earning differences between women and men, and daycare beyond 9 to 5.

- **Women of the World: Formal Laws and Policies Affecting Their Reproductive Lives**
  http://www.echonyc.com/~jmkm/wotw/

The Center for Reproductive Law & Policy, Inc. (CRLP) holds that "the promotion of women's reproductive rights is a crucial step toward the development of societies in which women hold equal status to men." This site focuses attention on six national governments and their formal laws, policies, and practices related to contraception, sterilization, abortion, population, and family planning. The six governments covered are those of Brazil, China, India, Germany, Nigeria, and the United States.

General Reference

- **Encyclopedia Smithsonian**
  http://www.si.edu/resource/faq/start.htm
    (see Encyclopedias Online)

- **Finding Data on the Internet: Links to Potential Story Data**
  http://nilesonline.com/data/
    This site is maintained by journalist Robert Niles with the intention of making it easy for other journalists to find statistics and data on the Internet. The following topics are covered: basic reference data, agriculture, aviation, banks and businesses, crime, economy and population, education, energy, finding people, health, immigration, law, military, nonprofits, politics, and weather. For each of these topics links to facts and information are provided. This site also provides links to the World Fact Book, the Library of Congress, and so on.

- **Free Internet Encyclopedia**
  http://clever.net/cam/encyclopedia.html
    (see Encyclopedias Online)

- **Reference Center of the Internet Public Library**
  http://ipl.sils.umich.edu/ref/RR/
    This site provides links to Internet resources on the following subjects: general reference, arts and humanities, business and economics, computers and the Internet, education, entertainment and leisure, health and medical sciences, law, government and political science, science and technology, and social sciences.

- **Usenet FAQs**
  http://www.cis.ohio-state.edu/hypertext/faq/usenet/
    (see FAQs)

Glass Ceiling

- **The Glass Ceiling**
  http://www.inform.umd.edu/EdRes/Topic/WomensStudies/ GenderIssues/GlassCeiling/MSPBReport/

This Web site contains the *U.S. Glass Ceiling Report,* more formally known as *A Question of Equity: Women and the Glass Ceiling in the Federal Government,* submitted to the president of the United States and Congress. The report asks why there are so few women in top-level civil service positions (1 in 10) when they make up 50 percent of the white-collar employees. The report identifies the barriers, unrelated to women's personal career decisions or qualifications, that play a role in this imbalance.

Graduate School

- **Graduate School Guide**
  http://www.schoolguides.com/
  This guide to graduate schools in the eastern and southern regions of the country allows users to search by field of study or state. E-mail addresses are given for those schools with graduate programs, as well as addresses and phone numbers.

- **Guide to Graduate School**
  http://www.jobtrak.com/gradschool_docs/gradschool/
  The UCLA Placement and Career Planning Center posts this site, which answers FAQs about graduate school. The menu includes the following topics: "Graduate School" (the basics), "Making the Decision," "Selecting a School," "Criteria for Evaluating a Graduate Program," "The Application Process" (which includes a personal essay and faculty recommendations), "Financing Graduate Studies," "Graduate School Tests," and "Reference/Resource Material."

- **Peterson's Education Center**
  http://www.petersons.com/
  (see Education)

Grammar

- **Elements of Style**
  http://www.columbia.edu/acis/bartleby/strunk/
  This is a guide to the proper use of the English language, with special focus on commonly misused words and expressions. It focuses on the rules of usage and writing principles most commonly violated.

- **Grammar Handbook**
  http://www.english.uiuc.edu/cws/wworkshop/writer.html
  Written by students at the University of Illinois, this handbook covers the parts of speech, phrases, clauses, sentences and sentence elements, and common usage problems. The section on common usage problems (when to use a colon instead of a semicolon,

where in a sentence to place modifiers) is especially useful for fine-tuning writing assignments.

Health Care and Health Services

- **Centers for Disease Control**
  http://www.cdc.gov/
  The CDC administers national programs aimed at the prevention and control of diseases, injury, and other preventable conditions. It develops and enacts programs to deal with environmental health problems such as chemical and radiation emergencies. It is also responsible for collecting and making available to the public national data on health status and health services.

- **Country Health Profiles for the Americas Only**
  http://www.paho.org/english/country.htm
  (see Pan American Health Organization)

- **Global Child Health News and Review Online Newspaper**
  http://edie.cprost.sfu.ca/gcnet/gchnr-01.html#News
  This news magazine covers issues related to children (such as health and well-being), as well as laws that directly affect families and children. Only the current issue is available at this site.

- **Health Care for Veterans**
  http://www.va.gov/medical.htm
  (see Veterans)

- **Health Status of the Elderly**
  http://aspe.os.dhhs.gov/96gb/bappend.txt
  (see Elderly)

- **International Red Cross/Red Crescent**
  http://www.ifrc.org/
  (see International Red Cross/Red Crescent)

- **World Health Organization Press Releases**
  http://www.who.ch/press/1996pres.htm
  The World Health Organization acts as " the directing and coordinating authority on international health work." Submit a keyword such as " malaria" or " Zaire" to see a list of all the press releases related to that disease or to health-related issues in that country. You can also choose to scroll through the list of press releases.

- **World Health Organization WWW Home Page**
  http://www.who.ch/
  (see World Health Organization)

Hispanics

- **Hispanic Online**
  http://www.hisp.com
  > *Hispanic Online* is a monthly magazine for and about Hispanics, covering events, issues, and news of interest to the Hispanic community. From this site you can access the latest issue and selected articles from back issues. This site also posts the "Hispanic 100," a list of the top 100 U.S. corporations that provide the most opportunities for Hispanics.

Historical Documents

- **Historical Letters, Documents, Essays, and Speeches**
  http://www.bungi.com/cfip/document.htm
  > The University of Kansas has posted an extensive list of historical documents beginning with Christopher Columbus's 1494 letter to the queen and king of Spain and ending with the 1993 Freedom of Information Act.

- **Historical Newspapers**
  http://lcweb.loc.gov/global/ncp/extnewsp.html#hist
  > This private collection of historical newspapers focuses on the early American experience, including the Colonial period, the Revolution, and the presidencies of Washington and Jefferson.

- **Historical Photographs**
  http://www.cmp.ucr.edu/
  > (see Visual Sociology)

- **Life History Manuscripts from the Folklore Project**
  http://lcweb2.loc.gov/wpaintro/wpahome.html
  > (see Life Histories)

- **List of All-American Memory Collection and Topics**
  http://lcweb2.loc.gov/ammem/amtitle.new.html
  > (see Visual Sociology)

- **National Archives and Records Administration**
  http://www.nara.gov/nara/whatsnew.html
  > The National Archives and Records Administration is an independent federal agency that maintains famous historical documents and enables people to view these documents. This site provides links to new NARA exhibitions and the Federal Register.

- **Search the White House Virtual Library**
  http://www.whitehouse.gov/WH/html/library-plain.html

This site allows users to search the White House databases for White House documents, radio addresses of the president, executive orders, and White House photographs. It is also possible to browse some historical national documents.

- **U.S. Historical Documents**
  gopher://wiretap.spies.com/11/Gov/US-History
  This site contains the text of a number of important documents beginning with the Declaration of Arms in 1775 and ending with the U.S. State Department's release of the text of the agreements reached at the 1944 Yalta Conference attended by Roosevelt, Churchill, and Stalin (see "World War II Documents").

Homeless

- **Homeless Fact Sheets**
  http://nch.ari.net/facts.html
  The Homeless Information Exchange and the National Coalition for the Homeless present a series of fact sheets on homelessness, such as "How Many Homeless People Are There?" and "Homeless Families with Children." Each sheet answers questions and includes a list of recommended reading.

Housing

- **Department of Housing and Urban Development**
  http://www.hud.gov/
  HUD is the federal agency responsible for developing programs to meet housing needs, ensure equal housing opportunities, and improve and
  develop existing housing.

Human Rights

- **Amnesty International Home Page**
  http://www.oneworld.org/amnesty/index.html
  (see Amnesty International)

- **Amnesty International UK Press Releases**
  http://www.oneworld.org/amnesty/ai_press.html
  This site gives information about refugees, activists, and political prisoners in other countries; on tourist safety abroad; and on human rights abuses in the United States and around the world. Press releases from the past are located at http://www.oneworld.org/amnesty/ ai_press_archive.html

- **Carter Center**
  http://www1.cc.emory.edu/CARTER_CENTER/
  (see Carter Center)

- **Country Reports on Human Rights Practices**
  http://homer.louisville.edu/groups/library-www/ekstrom/
  govpubs/goodsources/hrightsreports.html
  > This site contains the 1993, 1994, and 1995 U.S. Department of
  > State *Country Reports on Human Rights Practices*. For each
  > country (including the United States) you will find the State
  > Department's assessment of the political situation and human
  > rights policies, as well as a report on specific human rights abuses.

- **Universal Declaration of Human Rights**
  http://www.amnesty.org/aboutai/udhr.htm
  > The Universal Declaration of Human Rights was "adopted and
  > proclaimed by General Assembly Resolution 217 A (III) of 10
  > December 1948." This site offers the full text of this declaration,
  > which is put forth "as a common standard of achievement for all
  > peoples and all nations."

Immigration

- **U.S. Immigration & Naturalization Home Page**
  http://www.usdoj.gov/ins/
  > This Justice Department report provides information on legal
  > immigration to the United States during 1994. It includes the text
  > of the Immigration Act of 1990, and it reports on the number of
  > immigrants entering the United States under the act in 1992 and
  > 1993. The report also gives information such as the age, sex,
  > occupation, and place of intended residence for legal immigrants.

Information Infrastructure and Technology

- **Government and the Information Superhighway**
  http://www.nla.gov.au/lis/govnii.html
  > This site is maintained by the National Library of Australia and
  > provides links to policy statements by various governments
  > regarding national and global networking.

- **Information Infrastructure Task Force**
  http://www.iitf.nist.gov
  > The goal of the president's Information Infrastructure Task Force
  > (IITF) is to articulate and implement a vision for the national
  > information infrastructure. Members of the task force are high-
  > level representatives of the federal agencies that will play a major
  > role in shaping information and telecommunications technologies
  > and applications. The IITF posts a fact sheet, an organization chart,
  > a list of goals, publications, speeches, and activity reports.

- **Microsoft Press Releases**
  http://www.microsoft.com/corpinfo/prSearch.htm

Microsoft, world's leading software provider, has a goal of putting a computer on every desk in every home. Browse press releases from the past month, or search for press releases by month and year since 1994. Press releases deal with company information, new products, and services. Although this site is primarily a public relations tool for Microsoft, it provides useful information on technology and industry trends.

International Agencies

• **International Agencies and Information**
  http://www.lib.umich.edu/libhome/Documents.center/intl.html
  This site is an alphabetical list of international agencies beginning with the Asian Development Bank and ending with the World Trade Organization. Web links to these agencies and the information each agency offers is at one's fingertips. There are also links to the full text of many international treaties.

International Labor Organization

• **International Labor Organization**
  http://www.ilo.org
  The International Labor Organization determines international labor regulations such as the minimum standards of labor rights, working conditions, occupational safety standards and so on.

International Monetary Fund

• **The International Monetary Fund**
  http://www.self-gov.org/freeman/8904ewer.shtml
  The International Monetary Fund (IMF) was established to maintain fixed exchange rates among the different currencies of the world. The IMF does this by making short-term loans to nations with temporary balance-of-payment deficits.

International Red Cross/Red Crescent

• **International Red Cross/Red Crescent**
  http://www.ifrc.org/
  The International Red Cross cares for the sick and wounded in war and helps relieve suffering from pestilence, floods, fires, and other disasters. The hospitals, doctors, and nurses of the Red Cross remain neutral during war. The Red Crescent functions as the Red Cross in Turkey.

International Treaties

The sites listed below provide links to the full texts of dozens of international treaties and documents related to and explaining these treaties. Some of the treaties posted are GATT and NAFTA and treaties related to the nuclear testing, the environment, and much more.

- **Environmental Treaties and Resource Indicators**
  http://sedac.ciesin.org/pidb/pidb-home.html

- **International Treaties**
  gopher://gopher.nato.int/11/Information/treaties

- **Multilateral Treaties**
  http://www.tufts.edu/fletcher/multilaterals.html

Internet

- **CyberCourse**
  http://www.newbie.net/CyberCourse/00index0.html
  CyberCourse is posted by NewbieNET, an organization that provides free Internet training materials. CyberCourse teaches users to navigate the ever-changing and -expanding Internet. Each lesson is self-paced and designed to allow users an opportunity to sample the wide ranges of resources available on the Internet. The topics covered by CyberCourse include e-mail, Internet connections, list servers, search engines, urban legends, security, Archie, FTP, HTML, the World Wide Web, address searches, and so on.

- **CyberFriends**
  http://www.cyberfriends.com
  (see Pen Pals)

- **Declaration of the Rights of Netizens**
  http://www.columbia.edu/~rh120/netizen-rights.txt
  (see Netizens)

- **Emoticons and Abbreviations**
  http://pw2.netcom.com/~jampie96/emoticons.html
  (see Etiquette on the Net)

- **Internet Background**
  http://www.pbs.org/uti
  The Public Broadcasting Service has put together the "Beginners Guide to the Internet," which contains more than 300 links organized by topic to Internet-related sites. Everything you have ever wanted to know about the Internet but were afraid to ask can be found through this site. If you are a new user, read "Quick Tips" first. There is also an "Internet Quiz" to test your knowledge of the Internet.

- **Internet Terms**
  http://134.84.217.16/icec/training/terms.html
    (see Etiquette on the Net)

- **Netiquette**
  http://email.miningco.com/msub1.htm
    (see Etiquette on the Net)

- **Safety on the Information Highway**
  http://www.larrysworld.com/child_safety.html
    This site is posted by the National Center for Missing and
    Exploited Children and the Interactive Services Association.
    Although it is designed for parents in search of guidelines for their
    children's online computer use, this site is useful to anyone who is
    interested in a clear and balanced overview of the Internet and is
    concerned about online safety.

- **URL-Minder**
  http://www.netmind.com/URL-minder/URL-minder.html
    The NetMind Free Services page offers the URL-Minder. When
    you submit the URLs that are important to your research or
    personal interests, URL-Minder checks to see if the address has
    changed. If there is a change, it sends you an e-mail informing you
    of that change.

Journals and Newsletters Online

- **American Demographics Magazine**
  http://www.demographics.com/Publications/AD/Index.HTM
    American Demographics Magazine presents articles at this site
    dealing with demographic and marketing issues such as aging,
    racial classification, home workers, illegal immigration, consumer
    problems, and markets. You have the option of searching through
    each issue or of doing a keyword search.

- **CITYSCHOOLS**
  http://www.ncrel.org/ncrel/sdrs/cityschl.htm
    The first issue of the innovative journal *CITYSCHOOLS*, a
    research magazine about urban schools and communities, can be
    accessed at this site. *CITYSCHOOLS* rejects the "deficit model" as
    an approach to solving problems related to urban and inner-city
    schools and advocates a "resilience model" that emphasizes
    strengths.

- **Cover Stories from Previous Congressional Quarterly
    Weekly Report Stories**
  http://www.cq.com/wr.htm

From this site it's possible to read the cover stories from the *Congressional Quarterly Weekly* for any week in the previous year. The articles deal with political issues, most of which are in the news.

- **Disability Rag**
  http://www.etext.org/Politics/Disability.Rag/
  This site gives access to *The Disability Rag*, an online magazine that deals with issues disabled people face, such as access to public accommodations and discrimination in hiring practices.

- **Education Policy Analysis Archives**
  http://olam.ed.asu.edu/epaa/
  This site contains an electronically published scholarly journal focusing on education policy at all levels and in all countries.

- **Environmental Protection Agency (EPA) Journal**
  http://www.epa.gov/docs/epajrnal/
  The *EPA Journal* is published quarterly. The articles in each issue focus on an environmental theme such as Earth Day or environmental awareness.

- **Federal Bureau of Investigation (FBI) Law Enforcement Bulletin**
  http://www.fbi.gov/leb/leb.htm
  From this page it is possible to read past and present issues of the *Law Enforcement Bulletin,* a monthly publication by the FBI that centers around current trends and issues in law enforcement.

- **Foreign Service Journal**
  http://www.afsa.org/fsj/index.html
  The *Foreign Service Journal,* a monthly magazine of the American Foreign Service Association, is aimed at active and retired foreign service employees and at those interested in international diplomacy and U.S. foreign policy. Foreign affairs professionals, Foreign Service officers, and diplomatic correspondents write the articles for the journals. Some examples of titles of featured articles include "Has the U.S. Abandoned Its 'Best Friend' in Africa During 8-Year Civil War?" "Expanding U.S. Influence: After the Fall of USSR, America Moves Quickly to Establish Posts," and "Eulogy for a Consulate: Tiny U.S. Post Shuts Down in Bilbao After 2 Centuries of American Presence." Past and present issues are available for online viewing.

- **Global Child Health News and Review Online Newspaper**
  http://edie.cprost.sfu.ca/gcnet/gchnr-01.html#News

This news magazine covers issues related to children such as health and well-being, as well as laws that directly affect families and children. Only the current issue is available at this site.

- **Global Stewardship Network**
  http://www.iisd.ca/updates/
  This is a free news service covering issues related to global stewardship (that is, "caring for the earth and its current inhabitants, as well as a responsibility to leave to future generations a planet capable of sustaining life"). Past issues are also available at this site.

- **Hispanic Online**
  http://www.hisp.com
  Hispanic Online is a monthly magazine for and about Hispanics, covering events, issues, and news of interest to the Hispanic community. From this site you can access the latest issue and selected articles from back issues. This site also posts the "Hispanic 100," a list of the top 100 U.S. corporations that provide the most opportunities for Hispanics.

- **Indiana Journal of Global Legal Studies**
  http://www.law.indiana.edu/glsj/glsj.html
  From this site it is possible to read past and present issues of the *Indiana Journal of Global Legal Studies.* This interdisciplinary journal focuses on issues of global and local interest (the environment, AIDS, and so on), as well as markets, politics, technology, and culture.

- **Interracial Voice**
  http://www.webcom.com/%7Eintvoice/
  The *Interracial Voice* publishes articles that focus on the shortcomings of the U.S. racial classification system and that clarify the need for a new "racial" category for interracial individuals.

- **Journal of Statistics Education Information Service**
  gopher://jse.stat.ncsu.edu/1
  This site gives access to the *Journal of Statistics Education.* Many articles deal with subjects of interest to sociologists, and some articles represent especially good statistical analyses of social science data.

- **Multinational Monitor**
  http://www.essential.org/monitor/monitor.html
  Multinational Monitor is published bimonthly in January/February and July/August and once a month for the other eight months of

the year by Essential Information, an organization founded by Ralph Nader. *Multinational Monitor* covers corporate activity, especially in the "developing" world, as it relates to the environment, workers' health and safety, and labor union issues.

- **New Jour**
  http://gort.ucsd.edu/newjour/
  This site lists new journals and newsletters that have become available on the Internet in the past six months.

- **Newsprints**
  http://www.essential.org/newsprints/
  Newsprints is a bimonthly electronic journal put out by Essential Information, an organization founded by Ralph Nader. *Newsprints* scans more than one hundred of the highest-circulating daily regional newspapers for hard-hitting investigation of and commentary on issues of local, national, and international importance that the national media fails to cover. Note that the text of the articles and commentaries is not available online. *Newsprints* lists the information you need to track down the article: the title, the newspaper name, and the date it appeared.

- **Populi**
  gopher://fpa003.unfpa.org:70/11/populi
  Populi, a United Nations Population Fund publication, emphasizes issues related to overpopulation and fertility.

- **Postmodern Culture**
  gopher://jefferson.village.Virginia.EDU/11/pubs/pmc
  Postmodern Culture publishes interesting and creative work in the area of postmodernism.

- **Prison Legal News**
  http://www.etext.org/Politics/Prison.Legal.News/
  This site gives access to *Prison Legal News,* a monthly newsletter published by two prison inmates. The newsletter deals with court decisions and their effects on prisoners and families. The newsletters online are several years old, but they offer a unique perspective on prison issues.

- **Register of Leading Social Sciences Electronic Journals**
  http://www.clas.ufl.edu/users/gthursby/socsci/ejournal.htm
  This site compiles the leading online journals of value to researchers in the social sciences and humanities. Select the link "Alphabetical List of the Topics" and 19 screens of topics appear for you to choose from. Examples of topics are transportation, religion, and AIDS.

- **Scholarly Journals Distributed Via the Web**
  http://info.lib.uh.edu/wj/webjour.html
    University of Houston Libraries provides links to Web-based
    English-language scholarly journals that require no user
    registration fees

- **Society for the Study of Symbolic Interaction (SSSI): Papers of
  Interest**
  http://sun.soci.niu.edu/~sssi/papers/papers.html
    This site provides links to papers posted by the Society for the
    Study of Symbolic Interaction that represent good examples of
    research from a symbolic interactionist perspective.

- **Sociological Research Online**
  http://www.soc.surrey.ac.uk/socresonline/
    This page provides access to the latest issue of *Sociological
    Research Online,* an electronic journal that publishes "high quality
    applied sociology, focusing on theoretical, empirical, and
    methodological discussions which engage with current political,
    cultural, and intellectual topics
    and debates."

- **U.S. Department of Education Publications**
  http://www.ed.gov/publications.html
    This site is provided by the U.S. Department of Education and
    offers access to many of its publications, including newsletters,
    publications for parents, education statistics, and so on.

Kinship

- **The Genealogy Home Page**
  http://www.genhomepage.com/full.html
    Genealogy Roots Corner is the organization that sponsors The
    Genealogy Home Page. "The goal of Genealogy Roots Corner is
    to gather researchers together on one site" to share information
    that will help others searching for missing branches of the family
    tree and seeking to make family connections. This is an excellent
    comprehensive "library" to information on the Internet including
    (1) Genealogy Guides, (2) Libraries, (3) Maps, Geography, Deeds
    and Photography, (4) Newsgroups and Mailing Lists, (5)
    Genealogy Societies, (6) World Wide Genealogy Resources, and
    much more. There are more than 1600 links to Internet   resources.

- **WorldGenWeb**
  http://www.worldgenweb.org/

The WorldGenWeb Project is maintained in conjunction with   the
USGenWeb Project, a not-for-profit organization with   the
goal of making "genealogical research material available through
the Internet, ranging from various biographies to county vital
records (birth, death, marriage), etc." This site offers links to
Web pages offering genealogical information based in countries
ranging from Afghanistan to Zimbabwe.

Labor

• **Bureau of Labor Statistics News Releases**
  http://stats.bls.gov:80/newsrels.htm#OEUS
  " The Bureau of Labor Statistics (BLS) is the principal fact-finding
  agency for the Federal Government in the broad field of labor
  economics and statistics." Press releases cover the following broad
  list of topics: "Employment and Unemployment," "Prices and
  Living Conditions," "Compensation and Working Conditions,"
  "Productivity and Technology," "Employment Projections,"
  "International Comparison," and other BLS reports. Within each
  broad category are specific reports. For example, under the
  category "Compensation and Working Conditions" are news
  releases related to major work stoppages, occupational injuries,
  and union membership.

• **Department of Labor**
  http://www.dol.gov/
  The DOL enforces a variety of federal labor laws that guarantee
  workers' rights to safe and healthful working conditions, minimum
  hourly wages and overtime pay, freedom from employment
  discrimination, unemployment insurance, and workers'
  compensation.

• **Employment Statistics**
  http://www.bls.gov/cesprog.htm
  The Bureau of Labor Statistics provides information on the
  demographic makeup of the U.S. civilian work force. There is also
  information on the number of people employed by industry in
  1983 and 1994 with projections for the year 2005. In addition, the
  Bureau of Labor Statistics provides tables showing the 10
  industries with the largest projected job growth between 1994 and
  2005 and the 10 fastest growing occupations.

• **International Labor Organization**
  http://www.ilo.org
  (see International Labor Organization)

• **International Labor Organization Press Releases**
  http://www.ilo.org/public/english/235press/pr/index.htm

The International Labor Organization, a United Nations agency, seeks to promote social justice and establish internationally recognized standards of human and labor rights. Its press releases focus on labor issues such as child labor, unemployment, underemployment, equality for women, and international labor standards.

Law Enforcement

- **Federal Bureau of Investigation (FBI) Law Enforcement Bulletin**
  http://www.fbi.gov/leb/leb.htm
  From this page it is possible to read past and present issues of the *Law Enforcement Bulletin,* a monthly publication by the FBI that centers on current trends and issues in law enforcement.

Libraries

- **Library of Congress Home Page**
  http://lcweb.loc.gov/
  This is a source for links to publications, foreign and U.S. newspapers, the government, Congress, copyright laws and procedures, events and exhibits, and special collections.

- **Presidential Library System**
  http://www.nara.gov/nara/president/address.html
  The Presidential Library System includes nine presidential libraries and two presidential projects. These libraries are repositories dedicated to preserving and making available the papers, records, and other historical materials of past presidents, beginning with Herbert Hoover. There is information on museums and other public programs. The online library also offers biographical sketches, historical and personal information, political highlights, a genealogy, and an overview of the family life on each president.

- **State Libraries**
  http://www.dpi.state.wi.us/www/statelib.html
  The Wisconsin Department of Public Instruction posts this list of hypertext links to state libraries. Some of the libraries give only basic information such as their hours and phone numbers, but others offer access to library catalogs and resources, as well as convention and visitor's bureau information, state job bulletins, and legal information.

Life Histories

- **Life History Manuscripts from the Folklore Project**
  http://lcweb2.loc.gov/wpaintro/wpahome.html
  Life History Manuscripts from the Folklore Project, was a product of the Works Progress Administration (WPA). The federal government established the WPA in the 1930s to give employment

to those who could not find work during the Depression. Specifically the WPA Federal Writers' Project (1936-1940) gave work to unemployed writers or to anyone who could qualify as a writer. Approximately 300 writers from 24 states interviewed people across the country from all walks of life and circumstances. The product of their efforts was a collection containing 2,900 documents ranging from 2,000 to 15,000 words in length. The collection can be searched by keywords (such as textile workers, immigrants, or ex-slaves) and according to region or state.

Linguistics

- **Ethnologue: Languages of the World**
  http://www.sil.org/ethnologue/
  "*Ethnologue* is a catalogue of more than 6,700 languages spoken in 228 countries. The *Ethnologue Name Index* lists over 39,000 language names, dialect names, and alternate names. The *Ethnologue Language Family Index* organizes languages according to language families." In addition to the two indexes, the catalogue includes an introduction, which gives an overview of language and its importance to human interaction, and the   issues linguistics study.

- **The International Clearing House for Endangered Languages**
  http://www.tooyoo.L.u-tokyo.ac.jp/ichel.html
  The International Clearing House for Endangered Languages is part of the Department of Asian and Pacific Linguistics Institute of Cross-Cultural Studies at the University of Tokyo. The *ICHEL Newsletter* includes summaries of papers presented at the International Symposium on Endangered Languages (November 18-20, 1995) including "The Scope of Language Endangerment and Recent Responses to It*"* by Michael Krauss, "Minority Language Policy and Endangered Languages in China and Southeast Asia" by David Bradley, and "On Language Maintenance and Language Shift in Minority Languages of Thailand: A Case Study of So (Thavung)" by Suwilai Premsrirat. See also the "The Endangered Languages Project: A Progress Report" and "UNESCO Red Book On Endangered Languages."

- **Language in Cross-Cultural Understanding**
  http://www.halcyon.com/fkroger/bike/language.htm
  This essay by David Mozer emphasizes the importance of language for determining how we view the world. The author points out that biases are ingrained in our everyday language, and in order to accurately describe a culture, we must recognize

and alleviate biases through careful use of our words. The author also provides some tips on how to become "more sensitive, objective, and accurate in (our) observations of non-western cultures."

- **Linguistic Society of America**
  http://www.lsadc.org/
    The Linguistic Society of America, founded in 1924 and with a membership of 7,000, posts basic information about its organization along with the program of its annual meetings. Of particular interest to the linguistics student are the organization's statements and resolutions on issues such as "Language Rights," "Ebonics," "Research with Human Subjects," and "English Only Initiatives." The organization also posts of list of approximately 100 journals that publish articles related to linguistics. Finally, check out The *Field of Linguistics* a series of 22 essays explaining and clarifying the field of linguistics. Examples of essays include "Language Diversity," "Language and Brain," "Slips of the Tongue," "History of Linguistics," "Sociolinguistics," and "Endangered Languages."

Marx

- **Communist Manifesto**
  http://csf.Colorado.EDU/psn/marx/Archive/1848-CM
    The *Communist Manifesto*, an 1848 pamphlet, by Karl Marx and Friedrich Engels, outlines the principles of communism. "The Manifesto remains one of the world's greatest political documents, in content, style, and influence."

Maps

- **Interactive Atlas**
  http://www.mapquest.com/cgi-bin/
    mqatlas?screen=mqhome&link=wm_main&uid=80aa9pb4e08r
    pst
    Mapquest posts the Interactive Atlas, an online service that allows users access to county-level and city-level maps from over six continents. This atlas allows you to enter a street address along with its city, state, and zip code and to access a map of that address and surrounding streets and landmarks. A one-time free registration is required, which allows you to save and store maps for later use. A graphical browser is required to access this site.

- **Maps in the News**
  http://www-map.lib.umn.edu/news.html

The John R. Borchert Library of the University of Minnesota posts this site, which provides links to maps of regions that are in the news because of conflict or disaster. The number and quality of maps available varies by region. A graphical browser is required to access this site.

Media Contacts
- **Media List**
  http://link.tsl.state.tx.us/.dir/ready.dir/.files/medialist
  This site provides a list of e-mail addresses for newspapers, magazines, TV stations, and other media outlets that accept electronic submissions. This is a convenient method for writing letters or contacting media services with comments or for subscription information. Foreign news services are represented here as well.

- **National Press Club 1997-1998 Directory of News Sources**
  http://npc.press.org/sources/
  This National Press Club Web site offers a directory of news sources, including links to various agencies, magazines, and schools. It is designed for reporters and editors who need to find background information quickly for a story and the names of key contacts. The resources are grouped into subjects that begin with "Abortion/Reproductive Rights" and end with "Workplace Trends and Issues."

Media-Watch Resources
- **Media Watchdog**
  http://www.ipl.org/cgi-bin/redirect?http://theory.lcs.mit.edu/~mernst/media
  This site compiles online media-watch resources, including organizations, articles, censorship material, and other resources. The "Top Censored Stories" section is a good source of information on issues such as child labor, nuclear weapons, and the Internet.

Metric System
- **Mathematical Notation, Weights, and Measures**
  http://www.odci.gov/cia/publications/factbook/app-frame.html
  Do you need help converting liters into quarts or meters into yards? The CIA has posted conversion tables that list almost every metric measure and its English-system equivalent.

Movies
- **Movie Database**
  http://us.imdb.com/tour.html

This Web site defies brief description. We suggest that you take the Internet Movie Database Tour, which will give you a feel for the wide range of information housed at this site. There are more than a million entries covering 300,000 people, from actors/actresses to sound recording directors. There are plot summaries, FAQs, well-known quotations, trivia, and much more.

Multinational Corporations

- **Multinational Monitor**
  http://www.essential.org/monitor/monitor.html
  Multinational Monitor is published bimonthly in January/February and July/August and once a month for the other eight months of the year by Essential Information, an organization founded by Ralph Nader. *Multinational Monitor* covers corporate activity, especially in the "developing" world, as it relates to the environment, workers' health and safety, and labor union issues.

Multiracial People

- **Interracial Voice**
  http://www.webcom.com/%7Eintvoice/
  The *Interracial Voice* publishes articles that focus on the shortcomings of the U.S. racial classification system and that clarify the need for a new "racial" category for interracial individuals.

Music

- **HitsWorld**
  http://www.hitsworld.com/
  HitsWorld is a Web site that devotes coverage to top 5 and top 100 music charts. The site includes album reviews (and invites interested parties to submit reviews), Internet top 30 lists, and international charts.

National Debt

- **U.S. National Debt Clock**
  http://www.brillig.com/debt_clock/
  Debt clock maintainer Ed Hall posts this up-to-date report on the national debt. This page gives access to other sites concerned with the national debt and also answers some commonly asked questions about the national debt.

Native Americans

- **Bureau of Indian Affairs Press Releases**
  http://www.doi.gov/bia/press/index.html
  This site gives access to recent press releases from the Bureau of Indian Affairs. Court decisions and their impact on Native American communities are the most common subjects covered.

- **List of Federally Recognized Tribes**
  http://www.afn.org/~native/tribesl.htm
  The Native American Information Resource Server posts a list of federally recognized tribes.

- **Native Web**
  http://www.nativeweb.org/
  Native Web is posted by the Global Affairs Institute of the Maxwell School and provides links to information about people from around the world who are classified as "native" by virtue of birth, by their governments, or by way of life. The purpose is "to provide a cyber-place for the Earth's indigenous peoples" to communicate with one another and the world about "literature and art, legal and economic issues, land claims, and new ventures in self determination."

- **Persons of Indian Ancestry**
  http://www.doi.gov/bia/ancestry/ancestry.html
  This site describes how the government defines "Indian ancestry."

- **Top 25 American Indian Tribes for the United States**
  http://www.census.gov/ftp/pub/population/socdemo/race/indian/ailang1.txt
  The racial statistics branch of the U.S. Census Bureau lists the top 25 American Indian tribes in the United States for 1980 and 1990. There is also a table that shows the percentage change in membership for each tribe between 1980 and 1990.

Netizens

- **Declaration of the Rights of Netizens**
  http://www.columbia.edu/~rh120/netizen-rights.txt
  A netizen is the citizen of the 'net. Michael and Ronda Hauben at Columbia University have drafted a declaration of the rights (and obligations) that characterize responsible use of the Internet and its resources. The aim of the declaration is to promote a collective/cooperative spirit and to encourage universal access.

Network News

- **CNN Interactive**
  http://www.cnn.com/

- **Network News**
  The following sites provide links to television networks. Daily programming as well as news highlights, can be found at these sites.

NBC
http://www.nbc.com/

ABC
http://www.abc.com/

CBS
http://www.cbs.com/

- **Public Broadcasting Service**
  http://www.pbs.org
    This is the home page of the Public Broadcasting Service. This is an amazing resource for in-depth coverage of headline news and social issues in general. When you need an interesting topic for a paper or essay, go here first for ideas. You will not be disappointed. The quality of the material on this site defies simple summary. Start with the "Online Newshour" and browse its "past programs" and "essays and dialogues."

Newspapers

- **CReAte Your Own Newspaper (CRAYON)**
  http://crayon.net/
    CRAYON is a service maintained by Pressence Incorporated for managing news. After completing a free registration, users select the news sources from which they want to draw information. News sources are available at the international (such as *This Week in Germany*), national (such as *USA Today Nationaline*), and local (such as *The Detroit News*) levels. Users may select information to be taken from specific sections such as Sports or Weather, of the papers they choose. Users assign a name and motto to their paper and are given a URL address where they can read their paper daily. A graphical browser is required to access this site.

- **Ecola's Newsstand**
  http://www.ecola.com/news/
    Ecola's Newsstand provides links to 1,811 (and counting) English-language newspapers, magazines, and computer-related publications from around the world.

- **Hot News/Hot Research**
  http://www.poynter.org/research/reshotres.htm
    (see Social Issues)

- **Los Angeles Times**
  http://www.latimes.com/

- **News from Reuters Online**
  http://www.yahoo.com/headlines/

- **New York Times**
  http://www.nytimes.com/
  > In order to read *The New York Times* online, you must register the first time you visit this site. At the moment it is free.

- **State News**
  http://www.usatoday.com/news/states/ns1.htm
  > *USA Today* presents a significant news event for each of the 50 states.

- **Washington Post**
  http://www.washingtonpost.com/

- **TimeDaily**
  http://www.pathfinder.com/@@TLDE1QYAxeeEZBze/time/daily

- **Today in History**
  http://www.historychannel.com/today/

- **This Week's Magazine**
  http://pathfinder.com/@@At7@ngYAxeegunTi/time/magazine/domestic/toc/latest.html

- **USA Today**
  http://www.usatoday.com/

North American Free Trade Agreement

- **NAFTA**
  http://www.sice.oas.org/trade/nafta/naftatce.stm
  > The North American Free Trade Agreement is a 2,000-page document outlining plans to eliminate (over a 15-year period) all trade barriers between the United States, Canada, and Mexico. A directory allows you to select and view sections of the document.

Occupations

- **Detailed Occupation Information by Sex and Sometimes Race**
  gopher://una.hh.lib.umich.edu:70/00/census/summaries/eeous
  > This United States Equal Opportunity File is a list of hundreds of occupations and the number of males and females employed in each occupation. The first column of figures represents males, and the second females.

Overpopulation

- **Populi**
  gopher://fpa003.unfpa.org:70/11/populi
  Populi, a United Nations Population Fund publication, emphasizes issues related to overpopulation and fertility.

Pan American Health Organization

- **Country Health Profiles for the Americas Only**
  http://www.paho.org/english/country.htm
  The Pan American Health Organization assesses the health situation in each country that is part of the Americas. For each country specific health problems (statistics and a brief overview), demographic characteristics (population size and distribution and age-specific population characteristics), and descriptions and statistics related to health services and resources are available.

Pen Pals

- **CyberFriends**
  http://www.cyberfriends.com
  The purpose of CyberFriends is to provide an easy way for people sharing similar interests and professions to meet on the Internet. You simply fill out an application, which asks approximately 18 optional and required questions (age, sex, nationality, hobbies, personal philosophy, and so on). In two or three weeks your application will be posted free of charge. In the meantime you can search CyberFriends listings for profiles and e-mail addresses for the names of people with whom you might like to correspond.

Periodic Table

- **Periodic Table of Elements**
  gopher://ucsbuxa.ucsb.edu:3001/11/.Sciences/.Chemistry/ .periodic.table
  Do you need to know the symbol for arsenic? Or the melting point for carbon? This periodic table of elements is taken from the CRC *Handbook of Chemistry and Physics* and Lange's *Handbook of Chemistry.* It contains a comprehensive list of elements and provides information on the number, symbol, name, weight, boiling point, melting point, heat vapor, heat fusion, electrical conduct, thermal conduct, specific heat, specific gravity, valence, and configuration for each one.

Population, Countries

- **International Demographic Data**
  http://www.census.gov/ftp/pub/ipc/www/idbsum.html
  This Census Bureau site includes data on the population size of every country and territory in the world for 1950, 1960, 1970,

1980, 1990, and 1991-1995. Population size is also projected to the year 2000, as is age-specific population size.

- **Population for the Countries of the World**
  gopher://gopher.undp.org:70/00/ungophers/popin/wdtrends/ pop1996.txt
  This site give population data for the countries of the world as of 1994.

Population, County

- **See If Your County Lost or Gained Population 1990–1995**
  http://www.usatoday.com/news/states/special/pop/npop000.htm
  This *USA Today* site posts information on changes in population size between 1990 and 1995 for every country in the United States. Click on the name of a state for a listing of every county and each county's 1990 census count, estimated 1995 population size, and the percentage change. There is also a link to the list of the top 50 counties ranked by population gains and the bottom 50 counties ranked by population loss.

- **County Population 1900–1990**
  http://www.census.gov/population/www/censusdata/ cencounts.html
  This page gives the population of all counties or county equivalents in all 50 states from 1900 to 1990.

Population, United States

- **National PopClock from the U.S. Bureau of the Census**
  http://www.census.gov/cgi-bin/popclock
  The resident population of the United States is projected to the day, hour, minute, and second.

Population, World

- **World PopClock from the U.S. Bureau of the Census**
  http://www.census.gov/cgi-bin/ipc/popclockw
  This site gives the total population of the world, projected to the day, hour, minute, and second.

- **World Population**
  http://www.prb.org/media/index.htm#worldem
  The Population Reference Bureau provides information related to population and population growth around the world.

- **World Population Milestones**
  gopher://gopher.undp.org:70/00/ungophers/popin/wdtrends/ mileston

Do you know when the world's population reached a billion people for the first time? Do you know how many years passed before there were 2 billion people? Do you know the projected year the world's population will reach 7 billion people? Beginning with 1804, the year the world's population reached a billion people, this United Nation's Web site shows the subsequent years in which the world's population increased (or is projected to increase) by another one billion up to 11 billion.

Postmodernism

- **Postmodern Culture**
  gopher://jefferson.village.Virginia.EDU/11/pubs/pmc
  Postmodern Culture publishes interesting and creative work in the area of postmodernism.

Poverty

- **Data on Poverty**
  http://aspe.os.dhhas.gov/96gb/happend.txt
  The tables in this site relate to poverty in the U.S. as calculated according to the official census definition of poverty. The tables show the population, the number of people living in poverty, and the poverty rate in 1992 by age, race, region, and family type. For information on the official definition of poverty, see http://www.census.gov/cgi-bin/ print_hit_bold.pl/pub/hhes/www/povmea.html?poverty+threshold# first_hit

- **Federal Poverty Guidelines**
  http://aspe.os.dhhs.gov/poverty/poverty.htm
  Each year the department of Health and Human Services (HHS) issues poverty guidelines. The guidelines are used to determine financial eligibility for various federal programs such as Head Start, the Food Stamp Program, the National School Lunch Program, and the Low-Income Home Energy Assistance Program. At this site you can find the poverty thresholds (or the annual income level that marks the point at which families are considered living in poverty) for Alaska, Hawaii, and the 48 contiguous states and Washington, DC.

- **Poverty Clock**
  http://www.undp.org/undp/poverty/clock.htm
  This page defines poverty from a global perspective. It documents the number of people who are living on less than a dollar a day around the world and calculates the increase in poverty that occurs every minute. The "clock" can be viewed only on Netscape, but the information is available in a text format.

- **Where the American Public Would Set the Poverty Line**
  http://www.cdinet.com/Rockefeller/Briefs/
  This report is a Rockefeller Foundation *Research Brief on Poverty.*
  The importance of this report is that it points out the large
  discrepancy between where the public would set the poverty line
  and the official definition of poverty. The authors speculate on the
  policy and social implications of the discrepancy.

Press Releases and Briefings

Where does the media gets its information? Much of the
information and ideas for stories come in the form of press
releases. Check out these sites for firsthand accounts of
information released to the media.

- **Administration for Children and Families Press Releases**
  http://www.acf.dhhs.gov/ACFNews/press/
  This site is maintained by the Department of Health and Human
  Services. These press releases offer statistics on children and
  information about welfare demonstrations and reform.

- **Amnesty International UK Press Releases**
  http://www.oneworld.org/amnesty/ai_press.html
  This site gives information about refugees, activists, and political
  prisoners in other countries; on tourist safety abroad; and on
  human rights abuses in the United States and around the world.
  Press releases from the past are located at
  http://www.oneworld.org/amnesty/ai_press_archive.html

- **Bureau of Indian Affairs Press Releases**
  http://www.doi.gov/bia/press/index.html
  This site gives access to recent press releases from the Bureau of
  Indian Affairs. Court decisions and their impact on Native
  American communities are the most common subjects covered.

- **Bureau of Labor Statistics News Releases**
  http://stats.bls.gov:80/newsrels.htm#OEUS
  " The Bureau of Labor Statistics (BLS) is the principal fact-finding
  agency for the Federal Government in the broad field of labor
  economics and statistics." Press releases cover the following broad
  list of topics: " Employment and Unemployment," " Prices and
  Living Conditions," " Compensation and Working Conditions,"
  " Productivity and Technology," " Employment Projections,"
  " International Comparison," and other BLS reports. Within each
  broad category are specific reports. For example, under the
  category " Compensation and Working Conditions" are news
  releases related to major work stoppages, occupational injuries,
  and union membership.

- **Census Bureau Press Releases**
  http://www.census.gov/Press-Release/www
  The press releases at this site can be searched by subject or by
  date. Many of the press releases announce new statistics related to
  a wide range of subjects, from aging to national population
  estimates.

- **Census 2000 Press Releases**
  http://www.census.gov/Press-Release/www/2000.html
  This Web site contains Census Bureau press releases relating to
  Census 2000. Press releases announce plans for making the census
  simpler, less expensive, and more accurate and summarize
  recommendations made by various advisory committees. Most of
  the press releases relate to the issue of racial and ethnic
  classification and to plans to reach populations that have
  traditionally been undercounted.

- **Department of Veteran Affairs Press Releases**
  http://www.va.gov/pressrel/index.htm
  This site provides links to press releases covering U.S.
  veterans, including veterans' health issues (such as Gulf War
  syndrome, Agent Orange exposure, and health-care
  improvements), celebrations (Memorial Day, Veteran's
  Wheelchair Games), new programs, activities, and
  publications.

- **International Labor Organization Press Releases**
  http://www.ilo.org/english/235press/pr/index.htm
  The International Labor Organization, a United Nations agency,
  seeks to promote social justice and establish internationally
  recognized standards of human and labor rights. Its press releases
  focus on labor issues such as child labor, unemployment,
  underemployment, equality for women, and international labor
  standards.

- **International Monetary Fund (IMF) Press Releases**
  http://www.imf.org/external/news.htm
  The 181 countries that belong to the International Monetary Fund
  have pledged to cooperate with one another to maintain a
  productive and stable world economic environment. Members
  make monetary contributions from which "all may borrow for a
  short time to tide them over periods of difficulty in meeting their
  international obligations." IMF press releases announce the credit
  and loans that it has approved.

- **Microsoft Press Releases**

http://www.microsoft.com/corpinfo/prSearch.htm

Microsoft, the world's leading software provider, has a goal of putting a computer on every desk in every home. Browse press releases from the past month, or search for press releases by month and year since 1994. Press releases deal with company information, new products, and services. Although this site is primarily a public relations tool for Microsoft, it provides useful information on technology and industry trends.

- **NATO Press Releases**
http://www.nato.int/docu/pr/pr96e.htm

The North Atlantic Treaty Organization (NATO) is an alliance for collective defense formed in 1949 and linking 14 European countries, the United States, and Canada. Since 1989 (the fall of the Berlin Wall and the symbolic fall of communism) NATO has worked to establish cooperation between the governments of Central and Eastern European countries and with the newly independent states of the former Soviet Union. Its press releases focus on activities related to this task.

- **Organization for Economic Cooperation and Development (OECD) Press Releases**
http://www.oecdwash.org/PRESS/pr.htm

This Organization for Economic Cooperation and Development site provides access to press releases about OECD activities. The OECD is best known for its economic analyses and forecasts and for its advice to governments in the area of finance, investments, and job growth strategies. The *OECD Newsletter* is also posted on the Web site.

- **Population Reference Bureau Releases**
http://www.prb.org/media/pressrel.htm

The Population Reference Bureau's press releases focus on population-related issues (births, deaths, migrations) that affect life in the United States.

- **Today's Press Releases from the White House**
http://library.whitehouse.gov/PressReleases-plain.cgi

If you want to know the official opinion of the White House on a subject in the news, this is the site for you. This site contains all of the press releases coming directly from the White House for the current day. The press releases typically are related to speeches made by the president, first lady, and members of the White House staff or to information about them. The site also provides access to yesterday's press releases. In addition to White House press releases, there are also daily press briefings from the White House

press secretary. They can be found at
http://library.whitehouse.gov/Briefings-plain.cgi

- **United Nations Daily Press Briefings Search**
  http://www.un.org/News/briefings/
  UN press releases from the past five days can be accessed at this Web site. Past press releases can be searched by subject. Press briefings cover the UN's internal operations, as well as issues related to international peace, security, and cooperation.

- **U.S. International Trade Statistics Current and Past Press Release File**
  http://www.census.gov/foreign-trade/www/press.html
  These press releases deal with all aspects of trade (the movement of goods and services) between the United States and foreign countries.

- **World Bank Press Releases**
  http://www.worldbank.org/html/extdr/extme/press.htm
  These press releases report on World Bank loans, the conditions under which the loans were issued, and the effect of the loans on some segment of society.

- **World Health Organization Press Releases**
  http://www.who.ch/press/1996pres.htm
  The World Health Organization acts as "the directing and coordinating authority on international health work." Submit a keyword such as "malaria" or "Zaire" to see a list of all the press releases related to that disease or to health-related issues in that country. You can also choose to scroll through the list of press releases.

- **World Resources Institute News Release**
  http://www.wri.org/wri/press/wr96-nr.html
  "The mission of the World Resources Institute (WRI) is to move human society to live in ways that protect the Earth's environment and its capacity to provide for the needs and aspirations of current and future generations." The news releases for 1996–1997 focus on environmental imbalances associated with urbanization.

- **World Trade Organization (WTO) Press Releases**
  http://www.wto.org/press/press.htm
  The World Trade Organization, founded in 1995, resolves trade disputes, oversees trade policies, and facilitates trade negotiations. Its press releases report on WTO activities in these areas.

Prison Populations

- **Prison Legal News**
  gopher://gopher.etext.org/11/Politics/Prison.Legal.News
  This site gives access to *Prison Legal News,* a monthly newsletter published by two prison inmates. The newsletter deals with court decisions and their effects on prisoners and families. The newsletters online are several years old, but they offer a unique perspective on prison issues.

- **Profile of Inmates in the United States, England, and Wales, 1991**
  ftp://www.ojp.usdoj.gov/bjs/abstract/walesus.htm
  The Bureau of Justice Statistics posts this site, which provides an overview and selected highlights from a study that compares inmate populations and correctional systems in the United States with those in England and Wales.

Public Broadcasting Service

- **Public Broadcasting Service**
  http://www.pbs.org
  This is the home page of the Public Broadcasting Service. This is an amazing resource for in-depth coverage of headline news and social issues in general. When you need an interesting topic for a paper or essay, go here first for ideas. You will not be disappointed. The quality of the material on this site defies simple summary. Start with the "Online Newshour" and browse its "past programs" and "essays and dialogues."

Quotations

- **Bartlett's Familiar Quotations**
  http://www.columbia.edu/acis/bartleby/bartlett/
  This is a great source of phrases for personal as well as academic situations. For example, if you need to say something insightful about love after a fight, Richard Edward's (circa 1523–1566) "The anger of lovers renews the strength of love" might help. If you need to write an essay on the value of learning Aeschylus' (525–456 B.C.) line "Learning is ever in the freshness of its youth, even for the old" is a good starting point.

- **The Quotations Page**
  http://www.starlingtech.com/quotes/
  Choose "quotations by topic" if you are looking for quotations related to a specific topic. Choose "really miscellaneous" if you are looking for the statements of a particular person. This site offers a number of other search categories including "recent quotes," "advice," "sarcasm," and so on. Look for sociological concepts such as conformity, family inequality, and labor in the

list of topics. See how closely sociological ideas correspond with the ideas conveyed in the quotations.

Race

• **Institute on Race and Poverty**
  http://www.umn.edu/irp/

  The Institute on Race and Poverty at the University of Minnesota Law School was "created to focus attention on the unique dynamic created by the intersection of racial segregation and poverty." There are several publications with the entire text on-line (l) Examining the Relationship Between Housing, Education, and Persistent Segregation (2) In Pursuit of a Dream Deferred: Linking Housing and Education and (3) Race and Poverty: Our Private Obsession, Our Public Sin. Reexamining Racial Identity in Poor Communities of Color.

• **Office of Environmental Justice**
  http://es.inel.gov/oeca/oej.html

  "The Office of Environmental Justice (OEJ) was established in November 1992 with a broad mandate to serve as a focal point for ensuring that communities comprised predominately of people of color or low income populations receive protection under environmental laws."

• **One Drop of Blood**
  http://www.afn.org/~dks/race/wright.html

  This essay by Lawrence Wright, which appeared in *The New Yorker* on July 24, 1994, explores the problems associated with the attempt to classify people into clear-cut racial categories.

• **Questions of Race and Hispanic Origin**
  http://www.census.gov/population/www/socdemo/
    96natcontentsurvey.html

  The Census Bureau has posted the "Findings on Questions of Race and Hispanic Origin from the 1996 National Content Survey." This report summarizes how the census data on race and ethnicity would be effected by:(1) adding a multiracial response category in the race question; (2) placing the Hispanic origin question immediately before the race question; and (3) combining both of these changes.

Racial Classification

- **Race and Ethnicity Standards for the Classification of Federal Data on Race and Ethnicity**
  http://www1.whitehouse.gov/WH/EOP/OMB/html/fedreg/
  race-ethnicity.html
  (see Statistical Policy Directive No. 15)

- **Race and Hispanic Origin Figures for 1980 and 1990 Census**
  gopher://una.hh.lib.umich.edu:70/00/census/summaries/us-st-race
  This series of tables shows the total number and percentage of people classified by race and Hispanic origin living in four geographic regions of the United States and in each of the 50 states. There are also tables showing changes in the number and percentage of people in each racial and ethnic category between 1980 and 1990.

Refugees

- **Refugees International**
  http://www.refintl.org/
  Refugees International, founded in 1979, is an independent organization based in Washington, D.C. "heavily reliant on the support of committed and concerned individuals." The organization seeks to give governments and the UN early warnings about mass exoduses of refugees and to mobilize them to take action. In the last four years, Refugees International has responded to approximately thirty mass exodus crises, including calls of "Kurds stranded along the mountainous Turkish border; Burmese forced to flee to Bangladesh; war victims in Bosnia; Africans fleeing strife and famine in Liberia, Ethiopia, and Somalia; and Rwandans surging into Tanzania and Zaire." At the Web site you can access background information on refugee crises by region of the world and the latest news and developments related to new and ongoing refugee crises.

- **United Nations High Commissioner for Refugees**
  http://www.ids.ac.uk/eldis/data/d022/e02218.html
  This site provides "reliable and current information and analysis on all aspects relating to refugees and displaced persons, including their countries of origin, legal instruments, human rights, minorities, situations of conflict, and conflict resolution."

Religion

- **World Council of Churches**
  http://www.wcc-coe.org/
  (see World Council of Churches)

- **World Scripture: A Comparative Anthology of Sacred Texts**
  http://www.rain.org/~origin/ws.html
    The International Religious Foundation is an organization
    dedicated to promoting world peace through interreligious
    dialogue and cooperation. Select "Introduction" for more
    information on the purpose of the project. The foundation has
    posted the scriptures of several of the major world religions
    according to how each religion views (1) ultimate reality, (2)
    divine law, truth, and cosmic principles, (3) the purpose of human
    life, (4) life beyond death, and (5) the human condition, as well as
    16 other topics.

Research Methods

- **Sociology Research Papers**
  http://www.trinity.edu/departments/soc_anthro/research.html
    This Web site posted by the Department of Sociology and
    Anthropology at Trinity University covers the design (and
    rationale) of a research paper (not a term paper) in which the goal
    is to specify and test hypotheses derived from theory and to build
    on others' findings.

Social Issues

- **Documents in the News**
  **http://www.lib.umich.edu/libhome/Documents.center/
  docnews.html**
    This frequently updated site contains government documents
    relating to current news events. The University of Michigan
    Documents Center reports on government actions that make the
    news and on official government responses to domestic and
    international events such as the Montana Freemen standoff and the
    bombing of the U.S. military compound in Saudi Arabia.
    Documents in the news are also available for 1995.

- **Foreign Service Journal**
  http://www.afsa.org/fsj/index.html
    The *Foreign Service Journal,* a monthly magazine of the
    American Foreign Service Association, is aimed at active and
    retired foreign service employees and at those interested in
    international diplomacy and U.S. foreign policy. Foreign affairs
    professionals, Foreign Service officers, and diplomatic
    correspondents write the articles for the journals. Some examples
    of titles of featured articles include "Has the U.S. Abandoned Its
    'Best Friend' in Africa During 8-Year Civil War?" "Expanding
    U.S. Influence: After the Fall of USSR, America Moves Quickly to
    Establish Posts," and "Eulogy for a Consulate: Tiny U.S. Post
    Shuts Down in Bilbao After 2 Centuries of American Presence."
    Past and present issues are available for online viewing.

- **Hot News/Hot Research**
  http://www.poynter.org/research/reshotres.htm
  > Do you find that some news stories are hard to understand because you lack background information? The Poynter Institute for Media Studies, a nonprofit, journalism school in St. Petersburg, Florida, has created a Web site to help you understand the context behind the headline news. It offers links to Internet sites that will broaden your knowledge on a topic. For example, suppose you wanted to go beyond the headlines for more information on the 1996 bomb explosion at the U.S. Air Force base in Saudi Arabia. The Poynter Institute suggests links that will take you to (1) President Clinton's remarks, (2) the U.S. Air Force Web site, (3) press releases from the Defense Department, (4) a map of Saudi Arabia, (5) the 1995 *World Factbook* for background material on Saudi Arabia, and (6) the Saudi Arabia embassy.

- **Indiana Journal of Global Legal Studies**
  http://www.law.indiana.edu/glsj/glsj.html
  > From this site it is possible to read past and present issues of the *Indiana Journal of Global Legal Studies.* This interdisciplinary journal focuses on issues of global and local interest (the environment, AIDS, and so on), as well as markets, politics, technology, and culture.

- **Issues Page**
  http://www.colorado.edu/conflict/Cases/crcindex.html
  > This site is sponsored by the Institute for Global Communications, an organization dedicated to expanding and inspiring social action. It covers issues from acid rain to youth. Publications, organizations, conferences, and information sites are among the resources you will find at this site.

- **Newsprints**
  http://www.essential.org/newsprints/
  > *Newsprints* is a bimonthly electronic journal put out by Essential Information, an organization founded by Ralph Nader. *Newsprints* scans more than one hundred of the highest-circulating daily regional newspapers for hard-hitting investigation of and commentary on issues of local, national, and international importance that the national media fails to cover. Note that the text of the articles and commentaries is not available online. *Newsprints* lists the information you need to track down the article: the title, the newspaper name, and the date it appeared.

- **OneWorld Online Home Page**
  http://www.oneworld.org/index.html

This site is described by its creators as the largest collection of multimedia materials (text, graphics, audio, and video) in the world related to the issues of development, the environment, and human rights. It provides access to many resources including articles, journals, and guides. Fifty-seven groups, including Amnesty International, the Save the Children Fund, and UNICEF contribute material to this site.

- **Progressive Sociologists Network**
  gopher://csf.colorado.EDU/11/psn
  The Progressive Sociologists Network (PSN) invites sociologists (faculty and students) from around the world to participate in a forum that meets to discuss current issues related to the working class, civil rights, women's rights, race and ethnicity, community development, justice, the environment, and so on. Choose "Read Me Contents" for a directory and description of the PSN archives, which include student papers, syllabi for sociology courses, and book reviews. Check out "Martha's Corner" for thought-provoking discussions on topics ranging from the politics of identity to environmentalism in the former Soviet Union. If you are looking for an idea for an assignment related to social issues, this site is an excellent starting point.

Social Statistics

- **Basic Statistics on HIV and AIDS**
  http://www.cdc.gov/nchstp/hiv_aids/stats.htm
  (see AIDS and HIV)

- **Data on the Net**
  http://odwin.ucsd.edu/jj/idata
  This site is maintained by the University of California San Diego. It contains descriptions and links to 204 Web sites with numeric data ready to download, 63 data archives, and 109 social science gateways to social science resources.

- **Employment Statistics**
  http://www.bls.gov/cesprog.htm
  (see Labor)

- **Journal of Statistics Education Information Service**
  gopher://jse.stat.ncsu.edu/1
  This site gives access to the *Journal of Statistics Education*. Many articles deal with subjects of interest to sociologists, and some articles represent especially good statistical analyses of social science data.

- **Social Statistics Briefing Room**
  http://www.whitehouse.gov/fsbr/ssbr.html
  > This page provides access to current federal social statistics such as violent crime measures, demography (income by race, population of the U.S.), education (literacy rates, full-time graduate students), and health (leading causes of death, cases of measles). It also provides links to information produced by a number of federal agencies such as the Bureau of Justice Statistics, Federal Bureau of Investigation, U.S. Census Bureau, Health Care Financing Administration, and National Center for Education Statistics.

- **Statistical Abstract Frequently Requested Tables**
  http://www.census.gov/stat_abstract/
  > The Census Bureau has posted the tables for which there is the most demand as measured by the number of requests and responses to user surveys. Topics range from crime and crime rates to U.S. exports and imports of merchandise.

- **Statistical Briefs**
  http://www.census.gov/ftp/pub/apsd/www/statbrief/
  > Statistical briefs are 2- to 4-page reports that summarize data from demographic surveys of the U.S. population and provide data on issues of public policy. These documents are in Adobe Acrobat's Portable Document Format (PDF). In order to view these files, you must have Netscape, and you will need Acrobat Reader, which is available for free from the Adobe web site at http://www.adobe.com/acrobat/

- **Statistical Resources on the Web**
  http://www.lib.umich.edu/libhome/Documents.center/stats.html
  > This site provides links to statistics related to a large number of subjects ranging from agriculture to weather and including statistics of interest to sociologists.

- **USA Statistics in Brief: Part 1**
  http://www.census.gov/statab/USAbrief/part1.txt
  > This site contains statistics on population, law enforcement, education, communications, transportation, and housing. The data is from 1992.

Sociological Associations

The following is a list of international, national, and regional sociological associations. Although each posts different information, you will frequently find information on student competitions, association activities and meetings, new publications of interest to sociologists, membership details, and links to sociological resources on the Internet.

- **International Sociological Association**
  http://www.ucm.es/OTROS/isa/

- **American Sociological Association**
  http://www.asanet.org/

- **Mid South Sociological Association**
  http://www.uakron.edu/hefe/mssapage.html

- **Pacific Sociology Association**
  http://www.csus.edu/psa/psa.html

- **Society for Applied Sociology**
  http://www.indiana.edu/~appsoc/

- **Society for the Study of Symbolic Interaction (SSSI): Papers of Interest**
  http://sun.soci.niu.edu/~sssi/

- **Southern Sociological Association**
  http://www.MsState.Edu/Org/SSS/sss.html

Sociological Journals

- **American Sociological Association Sponsored Journals**
  http://www.asanet.org/Pubs/pubs.htm
  The American Sociological Association sponsors eight journals: (1) *American Sociological Review,* (2) *Contemporary Sociology: A Journal of Reviews,* (3) *Journal of Health and Social Behavior,* (4) *Social Psychology Quarterly,* (5) *Sociological Methodology,* (6) *Sociological Theory,* (7) *Sociology of Education,* and (8) *Teaching Sociology.* For each journal the ASA posts its mission statement and a table of contents for current and upcoming issues.

- **Current Research in Social Psychology**
  http://www.uiowa.edu/~grpproc/crisp/crisp.html
  The free online journal *Current Research in Social Psychology* is sponsored by the Center for the Study of Group Processes at the University of Iowa. It publishes manuscripts covering all aspects of social psychology. Some recent titles include "Feminine Speech

in Homogeneous Gender Groups," and "The Coalition Structure of the Four-person Family."

- **Electronic Journal of Sociology**
  http://www.sociology.org/
  The *Electronic Journal of Sociology*, founded in 1994, is an international online journal that publishes scholarly research from a widerange of theoretical traditions and methodological approaches that make an original contribution to sociological knowledge. Because *EJS* is an electronic journal, it especially encourages submissions exploring the sociological significance of information technologies or incorporating hypertext or other hypermedia features. Some recent titles include "Gender Advertisements Revisited: A Visual Sociology Classic?" and "Cyber McCarthyism: Witch Hunts in the Living Room."

- **Social Sciences Japan**
  http://www.iss.u-tokyo.ac.jp
  Social Sciences Japan is a quarterly newsletter of the Information Center for Studies in Japanese Society; a division within the Institute of Social Science (ISS) at the University of Tokyo. The articles are written by ISS research staff and other researchers connected with the institute. "Social Science Japan provides 'information about information' rather than detailed research reports" with the purpose of informing overseas researchers about research in their field.

- **Sociological Research Online**
  http://www.soc.surrey.ac.uk/socresonline/
  This page provides access to the latest issue of *Sociological Research Online,* an electronic journal that publishes "high quality applied sociology, focusing on theoretical, empirical, and methodological discussions which engage with current political, cultural, and intellectual topics and debates."

- **The SocioWeb: Journals and Magazines**
  http://www.socioweb.com/~markbl/socioweb/journals/
  SocioWeb, an "independent guide to the Sociological resources available on the Internet," posts this site which provides links to 15 sociological journals online. Some of the journals listed are: *Annual Review of Sociology, Gray Areas Magazine, Journal of Developing Societies, Social Research Update,* and *Sociological Research Online.* Most of the journals post full-text articles online and solicit submissions. A few listed journals, however, only allow users to view abstracts of articles online.

Sociological Resources

- **Berkeley Sociology Center**
  gopher://infolib.lib.berkeley.edu/11/resdbs/soci
  > The Berkeley Sociology Center provides links to sociological journals and data archives, as well as the Emma Goldman Papers, which chronicle the life of Emma Goldman (1869–1940), a major figure in the history of American radicalism and feminism.

- **Carl**
  telenet://CARL.LIB.ASU.EDU:23/
  > Arizona State University offers UNCOVER, a periodical index and document delivery system. Login with the password CARL and select VT100 (if applicable) as your terminal type. Select UNCOVER and follow instructions to enter via open access. Once in UNCOVER you can search major and obscure journals and periodicals by subject, author, or journal title. UNCOVER claims that over 4,000 current citations are added daily. You have the option of setting up an account if you plan use the document delivery service. If you plan to use the service to search and then find the document at a local library you do not have to set up an account.

- **Dead Sociologlists' Society**
  http://www.runet.edu/~lridener/DSS/DEADSOC.HTML
  > Sociologist Larry Ridener at Radford University posts this site which focuses on the lives and works of 16 important sociological theorists. Find personal information, a summary of ideas, and links to online works by Comte, Martineau, Marx, Spencer, Durkheim, Simmel, Weber, Veblen, Addams, Cooley, Mead, Park, Thomas, Dubois, Pareto, and Soroki. Ridener also provides links to Internet sites for topics ranging from Age Inequality to Violence/Abuse.

- **General Social Survey Subject Index**
  http://www.icpsr.umich.edu/gss/subject/s-index.htm
  > "The National Opinion Research Center (NORC) and the Inter-university Consortium for Political and Social Research (ICPSR), with support from the National Science Foundation have developed a state-of-the-art Internet services for users of the General Social Survey (GSS)." This site asserts that it is "suitable for both the novice and the experienced user" and that it is "useful for both research and teaching." Users can find longitudinal tables presenting results on opinion polls asking

questions about subjects ranging from Abortion and Affirmative Action to World Events and World View.

- **Internet Resources for Sociology**
  http://www.brynmawr.edu/library/Docs/sociol.html
    The Brigham Young University Library's Information Network has created links to Internet resources of interest to sociologists. The resources available include sociology departments, net links for sociologists, the *Electronic Journal of Sociology*, and *Yahoo*.

- **Sociological Subject Areas**
  http://www.pscw.uva.nl/sociosite/TOPICS/index.html
    This site is posted by The SocioSite, a project from the Department of Sociology at the University of Amsterdam. The SocioSite claims to be "a comprehensive information system which is very easy to use . . . . It contains high quality resources and texts that can be used as wheels for the sociological mind." The Sociological Subject Areas page provides links to approximately 150 sociological subject areas ranging from Activism to Youth. Each subject area contains links to articles, discussion groups, and reports from the United States and around the world.

- **Sociology Links**
  http://www.einet.net/galaxy/Social-Sciences/Sociology.html
    This site provides links to sociology departments, collections, directories, discussion groups, non-profit organizations, and periodicals from the United States and around the world.

- **Sociology Links at Princeton**
  http://www.princeton.edu/~sociolog/links.html
    Princeton University's Department of Sociology posts this site which provides links to Internet sites of interest to sociologists. The links at this site fall under the categories: Reference, Professional Associations, Institutions, Research Institutes, Centers, Sociological Practice, Web-Pages Devoted to Great Sociologists, Sociology Departments, Mailing Lists and Journals, Web-Pages of Paper Journals, University Press Catalogs, Data Archives, and Non-English Links.

- **Sociology Resources on the Internet**
  http://olympus.lang.arts.ualberta.ca:8010/vol001.003/
  nash.abstract.html

"Sociology Resources on the Internet" by Bradley Nash appeared in the August 1995 issue of the *Electronic Journal of Sociology*. Nash offers an introduction to selected sociology-related Internet resources including discussion groups (listservs), newsgroups, journals, associations, and virtual libraries.

- **The Best of 1995 Social Sciences & Humanities Internet Resources**
  http://coombs.anu.edu.au/SpcialProj/QLTY/BEST/Best96.html
  The Australian national University (specifically the Coombs Computing Unit, Research Schools of Social Sciences & Pacific and Asian Studies) sponsors this Best of the Web competition with the goal of discovering, acknowledging and publicizing the best Web resources in the Social Sciences, Humanities, and Pacific-Asian Studies. For the 1995 competition there were 80 nominations and 555 votes cast. The winning and nominated sites are posted. There is no doubt that participation will increase for future contests as the Web gains users. This Web site considers the questions "What constitutes quality on-line information?" and "What criteria should be used to evaluate Web sites?" The Coombs Computing Unit also posts a *What's New in WWW Social Sciences Online Newsletter* that can be accessed directly with the following URL:
  http://coombs.anu.edu.au/WWWVLPages/WhatsNewWWW/socsci-www-news.html.

- **The Sociology FAQ**
  http://www.sociology.net/socinfo/faq/
  The Sociology Corner answers nine frequently asked questions about sociology. Some of the questions addressed include: Who are important sociologists? What are the top thirteen books every sociologist should read? and, What kind of job can I get with a sociology degree?

- **World Lecture Hall-Sociology**
  http://www.utexas.edu/world/lecture/soc/
  This site is maintained by the World Lecture Hall which "contains links to pages created by faculty worldwide who are using the Web to deliver class materials." The site organizes the faculty webpages under headings and gives a brief description of each faculty page so users will know what to expect when accessing the page. Information is available on topics ranging from American Minorities to Women in East Asia.

Sociology Departments

- **Sociology Departments**
  http://www.socabs.org/html/16websa.htm
  This site provides links to sociology departments in the United States and around the world.

Sociology in Cyberspace

- **Michael Kearl's Home Page**
  http://www.trinity.edu/~mkearl/[ending here]
  Sociologist Michael Kearl at Trinity University is interested in cyberspace's potential "to inform and generate discourse, to truly be a 'theater of ideas.'" To demonstrate this potential Kearl has created more than 20 such "theaters," which explore topics of interest to any student of sociology. The topics and URL endings to the general address:
  http://www.trinity.edu/~mkearl/[insert ending here] are listed below.

Topic	Ending
The Times of Our Lives: Social Contours of the Fourth Dimension	time.html
A Sociological Social Psychology	socpsy.html
Social Gerontology	geron.html
Marriage and Family Life	family.html
Demography	demograp.html
Gender and Society	gender.html
Race and Ethnicity	race.html
Sociology of Death and Dying	death.html
Mass Media and Communication Studies	commun.html
Political Science	polisci.html
Health Statistics and the Medical Establishment	health.html
Anthropology	anthro.html
History	history.html
Science and Technology	science.html
Art and Art History	arthist.html
Paranormal Sites (with your mind in mind)	paranorm.html
Countdown	countdwn.html
Thinking About the Presidential Election	birthday.html
Lotto Mania	lotto.html
Credit Card Crazy	credcard.html
Reflections on the 40th Anniversary of the Television Remote Controller	remotetv.html

[We were so impressed with Professor Kearl's work we e-mailed him to ask about his future plans for these sites. His reply: "As is the case with this medium, the pages will be undergoing continuous updating and improvement. Given my teaching schedule for the year you may want to keep an eye on the social psych, times of our lives, and presidential election pages this fall; a new page on the sociology of knowledge will be brought on line in the spring. My goal is to move increasingly away from a listing of hypertext URLs and to instead weave them into overarching stories. See, for instance, the Countdown 2000 piece under Concluding on an Op-Ed Note."]

Space Exploration
- **National Aeronautics and Space Administration**
  http://www.nasa.gov/
  NASA is responsible for conducting research aimed at solving the problems of flight inside and outside Earth's atmosphere. It also determines the most effective use of the scientific and engineering resources of the United States with other nations involved with peacetime aeronautical and space activities.

State-Level Information
- **A Brief Guide to State Facts**
  http://phoenix.ans.se/freeweb/holly/state.htm
  This site includes basic information about each of the United States, including state capital, nickname, motto, flower, bird, tree, song, date entered the Union, and so on.

- **State and Local Governments**
  http://www.piperinfo.com/state/states.html
  This site provides detailed information from the states about their governments. The quantity and quality of the information varies by state, but often there is a link to the state constitution and to state-sponsored publications. In addition there is usually information on government services, including the office of tourism, libraries, archives, and public records. Also there are links to local, national, and international Web sites covering the state.

- **State Data**
  http://www.lib.virginia.edu/socsci/ccdb/state94.html
  The University of Virginia Social Sciences Data Center provides access to the 1994 *County and City Data Book.* Information such as the number of persons age 5 and over, percentage speaking a language other than English at home, number of votes cast for president in 1992, and so on is available for each of the 50 states. There are approximately 240 data tables from which to select.

- **State News**
  http://www.usatoday.com/news/states/ns1.htm

USA Today presents a significant news event for each of the 50 states.

- **State Profiles 1993—by the Small Business Association**
gopher://gopher.umsl.edu/11/library/govdocs/states
The Small Business Association compiles data from 15 different private and government sources to give a statistical overview of business activities in the states with emphasis on small business. This report includes general statistics about each state (top 5 industries, unemployment figures, number of business establishments, exports as a percentage of total U.S. exports). It also includes statistics related to small-business activity in each state (number of small businesses, small-business share of total employment, fastest-growing industries for small business, and so on).

- **State Rankings**
http://www.census.gov/ftp/pub/statab/ranks/
Drawing on data from the 1994 U.S. *Statistical Abstract*, this site ranks states according to 24 characteristics, including the value of exports, miles of motor vehicle travel, energy expenditures, and median household income. For more state rankings see http://www.census.gov/stat_abstract/ranks.html. This site contains rankings of the states according to things such as educational attributes, labor force composition, and number of motor vehicles.

- **U.S. State Fact Sheets**
http://www.econ.ag.gov/epubs/other/usfact/
This site, maintained by the U.S. Department of Agriculture, posts the most recent farm and rural data for each of the 50 states along with other general statistics about each state. The data available includes farm characteristics, farm financial indicators, top 5 agricultural exports, top 5 agricultural commodities, and top 5 counties in agricultural sales. This data is also available for the nation as a whole. This site is updated 3 or 4 times a year.

Statistical Policy Directive No. 15

- **Race and Ethnicity Standards for the Classification of Federal Data on Race and Ethnicity**
http://www1.whitehouse.gov/WH/EOP/OMB/html/fedreg/race-ethnicity.html
This file contains the text of *Statistical Policy Directive No 15: Race and Ethnic Standards for Federal Statistics and Administrative Reporting,* which has been in effect since 1977. It is the U.S. government's official policy on racial and ethnic classification. The file also contains an extensive critique of the

directive with suggestions for changing the system of classification.

Study Abroad

- **Cultural Immersion**
  http://www.nrcsa.com/
  The National Registration Center for Study Abroad gives information about immersion classes in 30 different countries. The online information covers program descriptions, dates, and fees.

- **Peterson's Education Center**
  http://www.petersons.com/
  (see Education)

- **Semester at Sea's Home Page**
  http://www.pitt.edu/~voyage/
  "The Semester at Sea is a floating university allowing students to experience diverse cultures while getting credit from the University of Pittsburgh." This page provides access to a general overview of the program, the mission statement, campus information, and information on the student body, enrollment, the academic program, the faculty, courses, and so on.

- **Study Abroad Home Page**
  http://www.studyabroad.com/
  This site is a resource for students to learn about study-abroad programs in 65 countries. Read the document "Consumer Information" before making any decisions about a program.

Supreme Court

- **U.S. Supreme Court Opinions**
  http://supct.law.cornell.edu/supct/
  Cornell University offers access to the text and analysis of Supreme Court opinions. You can access the most recent opinions (and dissenting opinions), as well as selected historic decisions. There is also biographical information on each of the Supreme Court justices.

Symbolic Interaction

- **Society for the Study of Symbolic Interaction (SSSI): Papers of Interest**
  http://sun.soci.niu.edu/~sssi/papers/papers.html
  This site provides links to papers posted by the Society for the Study of Symbolic Interaction that represent good examples of research from a symbolic interactionist perspective.

Tax Guides

- **Income Tax Information on the Internet**
  http://shell5.ba.best.com/%7Eftmexpat/html/taxsites.html
  TaxSites provides hypertext links to sites that have state and federal income tax–related information, including Usenet newsgroups and FAQs.

Thesaurus

- **ARTFL Project: Roget's Thesaurus Search Form**
  http://humanities.uchicago.edu/forms_unrest/ROGET.html
  A thesaurus is a dictionary that groups together words with similar meanings. This resource is helpful when you find yourself using the same word over and over.

Trade Practices

- **Country Reports on Economic Policy and Trade**
  http://www.state.gov/www/issues/economic/index.html
  This site provides access to the 1993 and 1994 *Country Reports on Economic Policy and Trade Practices*. You will find information on debt management policies, significant barriers to U.S. exports and investments, workers' rights, and much more.

- **World Trade Organization (WTO) Press Releases**
  http://www.wto.org/press/press.htm
  The World Trade Organization, founded in 1995, resolves trade disputes, oversees trade policies, and facilitates trade negotiations. Its press releases report on WTO activities in these areas.

Travel

- **City Destinations by Text Express**
  http://www.lonelyplanet.com/dest/text.htm#city
  (see Visual Sociology)

- **CityNet**
  http://www.city.net/
  This page is aimed at the traveler and offers information on 2,333 U.S. and international cities. It lists links to the most popular U.S. and international cities.

- **City Travel Guides**
  http://cityguide.lycos.com/index.html
  This travel guide provides links to cities, states, and regions that have posted information about a city in the United States or abroad. There is no telling what kinds of information you might find about a city. For example, under "Alexandria, Egypt" there is information about Cleopatra, Egyptian history, and Alexandrians

on the Internet. Under "Sidney, Australia" there is weather information, a picture gallery, and a virtual tour of the city.

- **Country Destinations by Text Express**
  http://www.lonelyplanet.com/dest/text.htm#count

  Designed for tourists, this site gives travel-related and background information on the countries of the world. It includes "Off the Beaten Path" and "Comments by Travelers" links. Travelers' comments are of sociological interest as they reflect the things tourists find important about their travel experience. Most comments focus on hotels and cuisine and indicate little interest in the lives and well-being of local peoples.

- **Travel Warnings and Consular Information Sheets**
  http://www.stolaf.edu/network/travel-advisories.html

  This U.S. Department of State site provides consular information sheets, travel warnings, and public announcements on every country in the world. The travel warnings are State Department recommendations to avoid travel to certain countries. Consular information sheets include factual information of interest to travelers, such as the location of the U.S. embassy or consulate, unusual entry regulations and immigration practices, health conditions, political disturbances, crime and security information, and drug penalties. Public announcements provide information about terrorist threats and other conditions the government believes pose risks to American travelers.

United Nations

- **United Nations Daily Press Briefings Search**
  http://www.un.org/News/briefings/

  UN press releases from the past five days can be accessed at this Web site. Past press releases can be searched by subject. Press briefings cover the UN's internal operations, as well as issues related to international peace, security, and cooperation.

- **United Nations Home Page**
  http://www.unicc.org/

  The United Nations, consisting of 126 countries, was formed to promote international peace, security, and cooperation.

United States Government Agencies

These are just a few of the government agencies represented on the Internet. The sites listed here may give access to government documents, newsletters, schedules of meetings, historical documents, statistics, and so on.

- **Centers for Disease Control**
  http://www.cdc.gov/

- **Consumer Information Center**
  http://www.pueblo.gsa.gov/textonly.htm

- **Department of Defense**
  http://www.dtic.dla.mil/defenselink

- **Department of Education**
  http://www.ed.gov/

- **Department of Housing and Urban Development**
  http://www.hud.gov/

- **Department of Justice**
  http://www.usdoj.gov/

- **Department of Labor**
  http://www.dol.gov/

- **Environmental Protection Agency**
  http://www.epa.gov/

- **National Aeronautics and Space Administration**
  http://www.nasa.gov/

- **U.S. Bureau of the Census**
  http://www.census.gov/

- **Federal Government Agencies**
  http://www.lib.lsu.edu/gov/fedgov.html
  This site gives access to hundreds of government agencies. Use it if you need to find an agency not listed previously.

- **Government Resources Via the Web—National, State, and Local**
  http://www.pic.net/dfwifma/govern.html
  This site provides access to 34 screens of available links to government documents. Documents are available from the following agencies or sources: U.S. federal government, U.S. Congress, Supreme Court, executive branch, Department of

Defense, Library of Congress, federal economy, governmental departments and agencies, and state and local governments.

Urban Schools

- **CITYSCHOOLS**
  http://www.ncrel.org/ncrel/sdrs/cityschl.htm
  The first issue of the innovative journal *CITYSCHOOLS,* a research magazine about urban schools and communities, can be accessed at this site. *CITYSCHOOLS* rejects the "deficit model" as an approach to solving problems related to urban and inner-city schools and advocates a "resilience model" that emphasizes strengths.

Urbanization

- **World Resources Institute**
  http://www.wri.org/wri/press/wr96-nr.html
  This site contains parts of the report *World Resources 1996–1997,* which gives statistical projections about urbanization through the year 2025. If you are interested in a particular issue of urbanization, you may conduct a keyword search. For example, entering the keyword "urban poor" produced definitions of "urban," information on China and India, and facts about population, poverty, and land degradation.

Veterans

- **Center for Women Veterans**
  http://www.va.gov/womenvet/CenWomVet.htm
  The Veterans Administration's Center for Women Veterans posts its mission statement and program goals, as well as historical information and statistics on the 1.2 million female veterans. There is also information on the unique health-care needs of female veterans, but especially sexual trauma.

- **Department of Veteran Affairs Press Releases**
  http://www.va.gov/pressrel/index.htm
  This site provides links to press releases covering U.S. veterans, including veterans' health issues (such as Gulf War syndrome, Agent Orange exposure, and health-care improvements), celebrations (Memorial Day, Veteran's Wheelchair Games), new programs, activities, and publications.

- **Veterans (Male and Female)**
  http://www.va.gov/vafvet.htm
  This is a statistical profile of the veteran population, including information on the number of veterans and the number of dependents and survivors of veterans

- **Health Care for Veterans**
  http://www.va.gov/medical.htm
  > This Veterans Administration site offers an overview of its health-care programs and facilities, which include 171 medical centers, over 362 outpatient-outreach facilities, 128 nursing homes, and 35 domicilaries. You can find information on the decentralized hospital computer system, telemedicine initiatives, and medical care cost recovery program. The VA also publishes an online collection of innovations in clinical practice with the goals of promoting information-sharing among clinicians and making a public statement about medical innovation within the VA system.

Violence

- **Statistics: The Juvenile Population in the U.S.**
  http://www.ncjfcj.unr.edu/homepage/g2.html
  > The University of Nevada at Reno hosts the Web site for the National Council for Juvenile and Family Court Judges. The document "Statistics: The Juvenile Population in the U.S." gives general characteristics of the juvenile population in the U.S. and statistics related in all types of violence directed toward self (suicide) and others.

Visual Sociology

- **AIDS Memorial Quilt Website**
  http://www.aidsquilt.org/
  > This AIDS Memorial Quilt Web site is posted by the Names Project, an organization with the goals of ending the AIDS epidemic and remembering those who have died from AIDS. There is information about the history of the quilt and the Names Project. Users are offered access to the quarterly newsletter, and they can also order Names Project memorabilia.

- **City Destinations by Text Express**
  http://www.lonelyplanet.com/dest/text.htm#city
  > This travel guide for selected cities around the world provides information about each city's history and hotel accommodations and includes links to other Web sites that focus on information about each city. Those with Netscape can view a slide show for each of the cities. The sociological significance of these slides lies with the images the tourist industry uses to attract people to a city.

- **Faces of Sorrow: Agony in the Former Yugoslavia**
  http://www.i3tele.com/photoperspectives/facesofsorrow/html/exhibition.html
  > Photo Perspectives sponsors this site, which attempts through a collection of 50 photographs to " give face to the faceless and

voice to the victims" of the 1991 war in Croatia and the four-year war in Bosnia and Herzegovina. Six sections ("Combatants," "Siege of Sarajevo," "Prisoners," "Refugees," "Faces of Rape," and "Ethnic Cleansing") depict the effects of this war on the faces of those caught in the middle.

- **Historical Photographs**
  http://www.cmp.ucr.edu/
  The California Museum of Photography posts this online historical photograph exhibition. The images include Pasadena in 1890, trains, logging, Ellis Island, Russia: before the revolution, Nagasaki, and Shirley Temple. The earliest photographs are from Japan and were taken in 1880.

- **International Museum of the Horse**
  http://www.imh.org/
  This museum portrays the historical relationship between people and horses and the functions to which the horse has been put in selected regions of the world. It includes a chronological overview, from an account of the first horse to today's racehorses.

- **List of All-American Memory Collection and Topics**
  http://lcweb2.loc.gov/ammem/amtitle.new.html
  The Library of Congress posts this site which consists of collections of primary source and archival material relating to U.S. culture and history, beginning with the Continental Congress in 1774 and ending with Carl Van Vechten photographs from 1964. There is access to six photographic collections, one recorded sound collection, five textual collections, and three early motion picture collections.

- **Portraits in Cyberspace**
  http://persona.www.media.mit.edu/1010/Exhibit/
  Portraits in Cyberspace is an online art exhibition hosted by the MIT Media Laboratory. There are 97 works in this exhibition. The artworks address the following: "Who inhabits the edges and margins of the online world? What constitutes identity in cyberspace—and how can it be portrayed? How are essential human experiences such as family, religion, community, sex, ethnicity, childhood, and personality being transformed in the digital era?"

- **Survivors: A New Vision of Endangered Wildlife**
  http://www.i3tele.com/photoperspectives/survivors/html/ survivors.html
  Photo Perspectives presents this collection of 37 portraits by James Balog of animals in danger of becoming extinct because of the

actions of humans. Balog challenges traditional images of animals in the wild as free to wander in nature as they please. Instead he portrays animals trapped in zoos and living in small spaces surrounded by modern cities and suburbs.

- **Tackiest Place in America Contest**
  http://www.thepoint.net/~usul/text/tacky.html
  The images registered at this site are the result of "The Tackiest Place in America Contest," sponsored by the Society for More Creative Speech. Places posted include "House Covered in Beer Cans" located in Houston, Texas; "The Running Fishboy" in Seiku, Washington; the Corn Palace in Mitchell, South Dakota; and a 60-foot statue of the Jolly Green Giant in Blue Earth, Minnesota. From a sociological point of view, this site is interesting because its "icons" speak to something of historical significance to that community even though it might be defined as tacky to an outsider (or insider for that matter).

- **Voyager's Interstellar Outreach Program**
  http://vraptor.jpl.nasa.gov/voyager/record.html
  On August 20 and September 5, 1977, the United States launched *Voyager I* and *Voyager II* probes into outer space to explore and photograph Jupiter, Saturn, Uranus, and Neptune. In 1990 the spacecraft left Earth's solar system. Attached to the outside of each was a gold-coated copper phonograph that contained 118 photographs of the planet and its inhabitants, 90 minutes of music from different eras and countries around the world, a collection of Earth's sounds, and greetings in 55 different languages. This site provides access to this "portfolio" of the planet, which was attached to the spacecraft "to send to any possible extraterrestrial auditors information about the Earth and its inhabitants." Imagine that you had the task of selecting items to represent life on Earth. What might you have done differently from the committee responsible for choosing its contents?

White House
- **Today's Press Releases from the White House**
  http://library.whitehouse.gov/PressReleases-plain.cgi
  If you want to know the official opinion of the White House on a subject in the news, this is the site for you. This site contains all the press releases coming directly from the White House for the current day. The press releases typically are related to speeches made by the president, first lady, and members of the White House staff or to information about them. The site also provides access to yesterday's press releases. In addition to White House press releases, there are also daily press briefings from the White House

press secretary. They can be found at
http://library.whitehouse.gov/Briefings-plain.cgi

Women

- **Center for Women Veterans**
  http://www.va.gov/womenvet/CenWomVet.htm
  (see Veterans)

- **Progress of Nations, 1996**
  http://www.unicef.org/pon96/contents.htm
  UNICEF posts the report *Progress of Nations, 1996*, containing articles, statistics, charts, and commentary on (1) women's physical well-being, (2) nutrition/malnutrition, (3) health, with emphasis on immunizations, (4) education of women, (5) the Convention on the Rights of Children, and (6) children's well-being in the industrial world.

- **The Glass Ceiling**
  **http://www.inform.umd.edu/EdRes/Topic/WomensStudies/
     GenderIssues/GlassCeiling/MSPBReport/**
  (see Glass Ceiling)

- **Statistical Indicators, UNICEF**
  http://www.unicef.org/pon96/statprof.htm
  (see Youth)

- **Women's Bureau**
  http://www.dol.gov/dol/wb/
  The Women's Bureau, an agency within the Department of Labor, formulates standards and policies aimed at improving wage-earning women's working conditions, advancing their opportunities for profitable employment, and informing them about employment rights and issues. From this site you can access press releases related to women's issues in the workplace, speeches, special reports, and other labor-related data. Some examples of special reports are *Working Women Count, Childcare Around the Clock,* and *Women's Bureau Fact Sheet*. Labor-related data can be found in "20 Facts on Working Women," "Earning Differences Between Men and Women," and "Women of Hispanic Origin in the Labor Force."

Work Abroad

- **PeaceCorps**
  http://www.peacecorps.gov/
  Peace Corps volunteers work in Africa, Asia-Pacific, Inter-America and the Caribbean, Eastern Europe, and the Mediterranean in the areas of agriculture, education, forestry,

health, engineering, skilled trades, business, the environment, urban planning, youth development, and the teaching of English for use in commerce and technology. This site gives information about the Peace Corps, becoming a Peace Corps volunteer, the places where Peace Corps volunteers work, and the Peace Corps global education program (which contains letters and interviews with Peace Corps volunteers).

- **Voluntary Service Overseas:**
  http://www.oneworld.org/vso/
  The Voluntary Service Overseas is an organization that recruits people between the ages of 20 and 70 to work in developing countries. This page answers questions about the program's goals, discuss volunteering, describes job openings, and gives general information about the program.

World as a Unit

- **1997 CIA World Factbook**
  http://www.odci.gov/cia/publications/factbook/
  This 1997 CIA World Factbook includes a section which considers the world as a unit. For example, it gives the unemployment rate, population size, total fertility, and so on for the world.

World Bank

- **The World Bank**
  http://www.worldbank.org/
  The World Bank is composed of five organizations that lend money to developing nations.

World Council of Churches

- **The World Council of Churches**
  http://www.wcc-coe.org/
  Nearly all Christian traditions are represented in the World Council of Churches, which is a fellowship of 330 churches from 120 countries in
  all continents.

World Economy

- **International Monetary Fund**
  http://www.self-gov.org/freeman/8904ewer.shtml
  (see International Monetary Fund)

- **International Monetary Fund (IMF) Press Releases**
  http://www.imf.org/external/news.htm
  The 181 countries that belong to the International Monetary Fund have pledged to cooperate with one another to maintain a productive and stable world economic environment. Members

make monetary contributions from which "all may borrow for a short time to tide them over periods of difficulty in meeting their international obligations." IMF press releases announce the credit and loans that it has approved.

- **Organization for Economic Cooperation and Development (OECD) Press Releases**
  http://www.oecdwash.org/PRESS/pr.htm
  This Organization for Economic Cooperation and Development site provides access to press releases about OECD activities. The OECD is best known for its economic analyses and forecasts and for its advice to governments in the area of finance, investments, and job growth strategies. The *OECD Newsletter* is also posted on the Web site.

- **World Bank**
  http://www.worldbank.org/
  (see World Bank)

- **World Bank Press Releases**
  http://www.worldbank.org/html/extdr/extme/press.htm
  These press releases report on World Bank loans, the conditions under which the loans were issued, and the effect of the loans on some segment of society.

- **World Trade Organization (WTO) Press Releases**
  http://www.wto.org/press/press.htm
  (see Trade Practices)

World Health Organization
- **World Health Organization WWW Home Page**
  http://www.who.ch/
  The World Health Organization strives to help people attain the highest level of health through technical projects and programs.

World Leaders
- **Heads of State and Heads of Government**
  http://www.geocities.com/Athens/1058/rulers.html
  This site lists all heads of state and other top-ranking leaders (past and present), including term of office and year of birth and death (if applicable), of all currently existing countries and territories. It also lists leaders (past and present) of the Arab League, European Union, Organization of African Unity, the Organization of American States, and the United Nations. A special section chronicles the changes in leadership since January 1996.

Writing Resources

- **On-Line Resources for Writers**
  http://www.ume.maine.edu/%7Ewcenter/resource.html
  This site gives links to a wide range of writing resources on the Internet including grammar, quotes, dictionaries, thesauruses, foreign-language dictionaries, citation guides, English as a second language, composition and rhetoric, and much more.

- **Purdue On-Line Writing Laboratory**
  http://owl.trc.purdue.edu/Files/Research-Papers.html
  The Purdue On-Line Writing Laboratory is a must for anyone writing a research paper. It takes you through the steps of writing a paper, from developing an outline to using statistics effectively. Paraphrasing, grammar, and citation formats are also covered at this site. There is even an exercise in which you can practice the fine art of paraphrasing and then check your work.

Youth

- **4-H**
  http://www.cyfernet.org/curricul/4hcommserv.html
  Statistics on the number of youth participating in 4-H programs and in specific 4-H projects and activities can be found at this site.

- **Statistical Indicators, UNICEF**
  http://www.unicef.org/pon96/statprof.htm
  UNICEF, a United Nations organization dedicated to children, publishes statistical indicators of child well-being, including the annual number of under-5 deaths, under-5 mortality rate, percentage of under-5 children underweight, percentage of children reading by grade five, and maternal mortality. This data is available for the countries in six major geographical area: (1) Sub-Saharan Africa, (2) the Middle East and North Africa, (3) Central Asia, (4) Asia and the Pacific, (5) the Americas, and (6) Europe.

Zipcode-Level Information

- **U.S. Gazetteer**
  http://www.census.gov/cgi-bin/gazetteer
  This Census Bureau site allows users to submit a zip code and/or the name of a city in the United States to get the population size and location in terms of latitude (position north or south of the equator measured from 0° to 90° and longitude (position to the east or west of an imaginary line running north and south around the world). Select "STF1A" or "STF3a" to choose among approximately 100 tables giving information on subjects ranging from age composition to the number of vacant housing units. If you have Netscape or another graphical browser, select "Map" to view a map of the city or zip code area.

- **United States Postal Service Zip Code Lookup and Address Information**
  http://www.usps.gov/ncsc/
  Submit a city name and its zip code appears. Submit a zip code and the name of the city to which that zip code has been assigned appears. The U.S. Postal Service answers commonly asked questions about zip codes and makes recommendations for addressing mail.

URLs for Specific Information

A African National Congress
gopher://gopher.anc.org.za
Document: ANC Home Page

AIDS
http://www.mit.edu/afs/athena/activity/m/medlinks/Resource-
Info/AIDS.intro
Document: An Introduction to AIDS

AIDS, underdiagnosing
http://www.indiana.edu/~aids/news/news1.html#hiv
Document: HIV Infection and AIDS in Rural America

http://www.aoa.dhhs.gov/aoa/pages/agepages/aids.html
Document: National Institute on Aging Age Page: HIV, AIDS, and
Older Adults

Alcoholics Anonymous, the BigBook of
http://www.recovery.org/aa/bigbook/ww
Document: The BigBook of Alcoholics Anonymous

B Banned books online
http://www.cs.cmu.edu/Web/People/spok/banned-books.html
Document: Banned Books On-line

Bhopal
http://www.essential.org/monitor/hyper/mm1294.html#ed
Document: Remembering Bhopal

Black plague
http://jefferson.village.virginia.edu/osheim/intro.html
Document: Plague and Public Health in Renaissance Europe

Body image, real and ideal
gopher://gopher.cc.columbia.edu:71/00/publications/women/wh27
Document: Body Image and Eating Disorders

C Capitalism
http://www.ocf.berkeley.edu/~shadab/capit-2.html
Document: Capitalism: Frequently Asked Questions (Theory)

Childcare, cost of
http://www.census.gov/population/socdemo/child/file1.txt
Document: Child Care Costs and Arrangements

Civil Right Act of 1991
http://www.hg.org/1215.txt
Document: The Civil Rights Act of 1991

Collective memory
http://www.saed.kent.edu/Architronic/v2n2/v2n2.05.html
Document: Ritual and Monument

Conservation efforts, water
http://www.uswaternews.com/archive/96/conserv/consort.html
Consortium Promotes Washing Machines That Use Less Water,
Energy

http://www.uswaternews.com/archive/96/conserv/swdrou.html
Document: Emergency Water Conservation Measures Implemented
in Drought-Stricken Southwest

http://webf0134.ntx.net/products/plubming/waterwise/index.html
Document: Be Water-Wise: 7 Ways in 7 Days

http://ianrwww.unl.edu/ianr/waterctr/wctriv.html
Document: Water Trivia

http://www.mbnet.mb.ca/wpgwater/welcome.html
Document: Welcome to the Waterfront—Winnipeg's Water
Conservation Information Source

http://www.uswaternews.com/archive/96/conserv/albuq.html
Document: Albuquerque Saves a Billion Gallons in '95 Usage

Correlation, causation, and prediction
http://www.stat.ncsu.edu/info/jse/v2n2/datasets.rossman.htm
Document: Televisions, Physicians, and Life Expectancy

Credit
http://www.lib.virginia.edu/journals/EH/EH37/Murphy.html
Document: The Advertising of Installment Credit

Cultural relativism, as applied to poopets
http://www.poopets.com/TEWKS.HTML
Document: Tewksbury Gardens

Culture
http://www.loc.gov/folklife/cwc.html
Document: American Folklife - A Commonwealth of Cultures

http://pantheon.cis.yale.edu/~rmelende/alexander.html
Document: Culture Symposium

Cultural issues (ideas for a paper on culture)
http://www2.uchicago.edu/jnl-pub-cult
Document: Public Culture Home Page

Cyberspace, Sociology of
http://www.bradley.edu/las/soc/syl/391/
Document: The Sociology of Cyberspace

D Demography
ftp://coombs.anu.edu.au/coombspapers/coombsarchives/demography/
what-is-demography.txt
Document: What is Demography?

http://epn.org/sage/rstill.html
Document: "Soft" Skills and Race: An Investigation of
Black Men's Employment

Domestic violence
http://www.ojp.usdoj.gov/bjs/abstract/spousmur.htm
Document: Husbands Convicted More Often than Wives for
Spouse Murder

http://www.ncjrs.org/txtfiles/against.txt
Document: Women Usually Victimized by Offenders They Know

Downsizing, effects on income
http://www.census.gov/ftp/pub/hhes/laborfor/dewb9092/jobturntab.html
Document: Distribution of Average Weekly Earnings Before and
After Job Turnover

Drop-out rate
gopher://INET.ed.gov:12002
Path: ERIC.src; enter ED386515
Document: School Dropouts: New Information about an Old
Problem

E   Electronic learners
http://albie.wcupa.edu/ttreadwell/brrec.html
Document: The Bill of Rights and Responsibilities for
Electronic Learners

Economy, information-based
http://www.educom.edu/web/pubs/review/reviewArticles/31144.ht
ml
Document: The Information Economy: How Much Will Two Bits
be Worth in the Future

Education, assessment
gopher://gopher.ed.gov:10000/00/tab/assess/naep/math/readme
Document: National Assessment of Educational Progress 1992
Mathematics Assessment

Education, funding
gopher://INET.ed.gov:12002
Path: ERIC.src; enter ED350717 (if number doesn't work, plug in
financial equity)
Document: Financial Equity in the Schools

Employment, equal opportunity
http://www.law.cornell.edu/uscode/42/2000e-2.html
Document: Unlawful Employment Practices

F   Family, coalition structures within
http://www.uiowa.edu/~grpproc/crisp/crisp.1.3.html
Document: The Coalition Structure of the Four-Person Family

Feminism
http://www.cis.ohio-state.edu/hypertext/faq/usenet/feminism/
info/faq.html
Document: SOC. Feminism Information

http://www.rochester.edu/SBA/declare.html
Document: The Declaration of Sentiments

Fetal Alcohol Syndrome
http://silk.nih.gov/silk/niaaa1/publication/aa13.htm
Document: Fetal Alcohol Syndrome

Foreign-born
http://www.census.gov/ftp/pub/population/socdemo/foreign/97/ppltab2.txt
Document: Foreign-Born Report Highlights

Foreign labor statistics
http://stats.bls.gov/flsfaqs.htm
Document: FLS Frequently Asked Questions

Foreign-owned manufacturing establishments in the USA
ftp://146.142.4.23/pub/news.release/History/fome.102793.news
Document: Foreign-owned Manufacturing Establishments in USA

Fundamentalism, religious
http://www.ais.org/~bsb/Herald/Previous/495/
fundamentalism.html
Document: What Does Fundamentalism Really Mean?

G Gallup Polls
http://www.gallup.com/
Document: Welcome To The Gallup Organization

Gangs, mothers against
http://www.winternet.com/%7Ejannmart/nkcmag.html
Document: Mothers Against Gangs

Gender, the fluidity of
http://ezinfo.ucs.indiana.edu/%7Emberz/ttt/articles/rights
Document: The International Bill of Gender Rights

Gender inequality, in wages and income
http://www.dol.gov/dol/wb/public/wb_pubs/wagegap2.htm
Document: Earnings Differences Between Women and Men

http://www.dol.gov/dol/wb/public/wb_pubs/wwmf1.htm
Document: Women Who Maintain Families

Goals 2000
http://commons.somewhere.com/reportcard/1995/
DAILY.REPORT.CARD113.html#Index2
Document: Goals 2000: A Politically Charged Battleground

gopher://gopher.ed.gov:10001/00/initiatives/goals/

legislation/g2k-1
Document: National Education Goals

H Hate crimes
http://www.getnet.com/silent
Path: Links/Hate Crimes Advisory Committee
Document: Phoenix Police Department: Hate

Healthcare, role of faith
http://www.interaccess.com/ihpnet
Path: /text-only home page/IHP-NET Documents
Archive/Expanding the Public Health Envelope Through Faith
Community
Document: Expanding the Public Health Envelope Through
Faith Community

High school graduates, college enrollment among
FTP://ftp.bls.gov/pub/news.release/hsgec.txt
Document: College Enrollment and Work Activity of 1995 High
School Graduates

Housing statistics
http://www.census.gov/ftp/pub/hhes/www/housing.html
Document: America's Housing

Hypertext
http://www.teleport.com/~cdeemer/essay.html
Document: What Is Hypertext?

http://www. rtvf.nwu.edu/telecine
Document: Turning the Page of Journalism

I Immigration, undocumented
http://www.law.indiana.edu/glsj/vol2/calavita.html
Document: U.S. Immigration Policy: Contradictions and
Projections for the Future

Income, operational definitions
http://www.census.gov/cdrom/lookup
Path: /STF3 technical documentation/definition of subject
characteristics/ poverty status in 1989

http://www.census.gov/td/stf3/append_b.html
Path: /Income in 1989

Independent Study
http://www.educom.edu/educom.review/review.95/jul.aug/
twigg.html
Document: The Value of Independent Study

Industrial revolution, effect on paper making
http://www.ipst.edu/amp/machine.html
Document: The Advent of the Paper Machine

Information infrastructure, global
http://www.ibm.com/ibm/gii/sectionI.html
Document: IBM and the Global Information Infrastructure

International Trade Lists
http://www.census.gov/ftp/pub/foreign-trade/www/top10.html
Document: Top 10 International Trade Lists

Internet
http://info.isoc.org/speeches/interop-tokyo.html
The Present and the Future of the Internet: Five Faces

http://info.isoc.org/papers/truth.html
Document: Truth and the Internet

ftp://ds.internic.net/rfc/rfc1087.txt
Document: Ethics and the Internet

http://www.educom.edu/web/pubs/review/reviewArticles/
31216.html
Document: Raison d'Net: Are Your Ready for the Thing Called
Change

Internet addiction, symptoms of
http://home.hkstar.com/~joewoo/symptom.html
Document: Symptoms of Internet Addiction

Internet, cemetery on
http://www.cemetery.org/
Document: World Wide Cemetery

Internet history
http://scuba.uwsuper.edu/~rwhiffen/web-
intro/history/welcome.html

Document: Ten Minutes of Internet History

Internet, its effects on society
http://sunsite.unc.edu/horizon/pastissues/vol2no5/lead.html
Document: Redefining Success: Public Education in the 21st
Century

http://www.educom.edu/web/pubs/review/reviewArticles/
31134.html
Document: Cyber-Cops: Angels on the Net

http://www.islandnet.com/%7Ercarr/oddy.html#mentornumbers
Document: Oddysey, Volume 3.2, May, 1995

http://www.lincoln.ac.nz/reg/futures/renaiss2.htm
Document: Renaissance Two: Second Coming of the Printing Press

Internment, WWII
http://www.netzone.com/~adjacobs/
Document: European American Internment

Intersexed
http://www.isna.org/FAQ.html
Document: Frequently Asked Questions

IRS - Internal Revenue Service
http://www.irs.ustreas.gov/plain/cover.html
Document: IRS - The Digital Daily

L   Literacy
    gopher://INET.ed.gov:12002
    Path: ERIC.src; enter ED372664
    Document: Estimating Literacy in the Multilingual U.S.

    Literacy, new forms of
    http://www.educom.edu/web/pubs/review/reviewArticles/
    31231.html
    Document: Information Literacy as a Liberal Art

M   Mandela's speech upon release from prison
    gopher://wiretap.spies.com/00/Gov/US-Speech/mandela
    Document: Nelson Mandela's Speech Upon Release

Migration
http://www.census.gov/population/socdemo/migration/tab-a-3.txt
Document: Annual Inmigration, Outmigration, and Net Migration
for Metropolitan Areas: 1985-1994

Migration, international (see foreign-born)
gopher://gopher.undp.org/00/ungophers/popin/wdtrends/inttab
Document: South-to-North Migration Flows

Misinformation, the role of television in
http://www.law.indiana.edu/fclj/pubs/v47/no1/jrhodes.html
Document: Even My Own Mother Couldn't Recognize Me

Mortality, infant and child
http://www.census.gov/ftp/pub/statab/ranks/pg05.txt
Document: Birth and Infant Mortality Rates in the United States by
State and Area in 1992

gopher://gopher.undp.org/00/ungophers/popin/wdtrends/child
Document: Child Mortality Estimates

Multinational corporations
http://www.essential.org/monitor
Document: Multinational Monitor

Multinational corporations, effect on community
http://www.igc.apc.org/elaw/update_summer_95.html#hyundai
Document: Hyundai Plans Chip Factory for Eugene

Multinational corporations, Fortune Magazine's most admired
http://pathfinder.com/@@oz00hwuapotknqeg/fortune/magazine/
specials/mostadmired/comebacks.html
Document: Fortune Magazine's Most Admired Corporations

Multinational corporations, worst
http://www.essential.org/monitor/hyper/mm1294.html
Document: The Ten Worst Corporations of 1994

N Netizens
   http://www.columbia.edu/~hauben/netbook/
   Netizens: An Anthology or Netizens: On the History and Impact of
   the Net

   Netizens, rights of

http://www.columbia.edu/~rh120/netizen-rights.txt
Document: Proposed Declaration of the Rights of Netizens

Nuclear testing
http://engagedpage.com/~engaged/hcant/
Document: ~A~ Hawaii Coalition Against Nuclear Testing

O Occupational structure, projected changes in
FTP://ftp.bls.gov/pub/news.release/ecopro.txt
Document: BLS Press Release New 1994-2005 Employment
Projections

P Poverty, federal guidelines
http://aspe.os.dhhs.gov/poverty/poverty.htm
Document: HHS Poverty Guidelines

Proposition 187
http://www.digitas.harvard.edu/~perspy/issues/1995/may/187cons.
html
Document: The Aftermath of Prop 187: Licensing Human Rights
Abuses Against Racial Minorities

R Racial Categories, personal experiences
http://www.webcom.com/intvoice/letters1.html
Document: Letters/Voices to the Editor

http://www.webcom.com/intvoice/letters3.html
Document: Letters/Voices to the Editor

Racially mixed people, bill of rights
http://www.webcom.com/intvoice/rights.html
Document: Bill of Rights for Racially Mixed People

Racist ideology, examples of
http://odur.let.rug.nl/~usa/P/tj3/writings/slavery.htm
Document: Thomas Jefferson On Slavery

http://www.halcyon.com/pub/FWDP/Americas/manifest.txt
Document: Reflections on Race and Manifest Destiny

Rape
gopher://justice2.usdoj.gov/00/fbi/January95/jan3.txt
Document: The Art of Interrogating Rapists

Research methods, ethics in
http://sun.soci.niu.edu/~sssi/papers/dirty.data
Document: Dirty Information and Clean Conscience:
Communication Problems in Studying 'Bad Guys'

Religion, media coverage of
http://www.missouri.edu/~c671545/journalism.html
Document: Arguments For and Against More Coverage of
Religion in the News Media

Research, guide to
http://www.dnai.com/~children/report_guide.html
Document: Step by Step Guide

Resilience model (versus deficit model)
http://www.ncrel.org/ncrel/sdrs/cityschl/city1_1b.htm
Document: Resilience Research: How Can It Help City Schools

http://www.ncrel.org/ncrel/sdrs/cityschl/city1_1c.htm
Document: Funds of Knowledge: A Look at Luis Moll's Research
Into Hidden Family Resources

Resilience Research (vs. deficit model)
http://www.ncrel.org/ncrel/sdrs/cityschl.htm
Document: Who Are Today's City Kids? Beyond the "Deficit
Model"

S   School prayer
http://www.religioustolerance.org/prayer.htm
Document: Prayer in Public School

Schools, rural
gopher://INET.ed.gov:12002
Path: ERIC.src; enter ED317332
Document: Small Schools: An International Overview

Secularization
http://www.Trinity.Edu/~mkearl/never.html
Document: You Never Have to Die!

http://www.interaccess.com/ihpnet/andrews
Document: The Forgotten Player in Health Care
Reform: Organized Religion

Selective Service System
http://www.sss.gov/
Document: Selective Service Home Page

Sex segregation
http://www.ul.cs.cmu.edu/books/sex_segregation/sex001.htm
Document: Sex Segregation in the Workplace

Social class
http://jse.stat.ncsu.edu:70/0/jse/v5n1/simonoff
Document: The "Unusual Episode" Data Revisited

Social mobility
http://www.census.gov/ftp/pub/hhes/housing/ahs/tab2-11.html
Document: Why Move?

Status, defined by neighborhood characteristics
http://propertyguide.com/steps.html#Step
Document: Step 2: Selecting a New Neighborhood

T   Television, violence on
    http://www.mediascope.org/ntvs1.htm
    Document: National Television Violence Study

    Tests, misuses of
    gopher://INET.ed.gov:12002
    Path: ERIC.src; enter ED315429
    Document: Five Common Misuses of Tests

    Travel and disease
    http://www.cdc.gov/ncidod/EID/vol1no2/wilson.htm
    Document: Travel and the Emergence of Infectious Diseases

U   Unabomber's Manifesto
    http://www.hotwired.com/special/unabom/list.html
    Document: HotWired: Unabomber's Manifesto: Index

V Valentine's Day
http://www.census.gov/Press-Release/fs98-02.html
Document: Census Bureau Facts for Valentine's Day

W Wages
FTP://ftp.bls.gov/pub/news.release/occomp.txt
Document: New Survey Reports on Wages and Benefits for
Temporary Help Services Workers

http://stats.bls.gov/news.release/annpay.nws.htm
Document: Average Annual Pay by State and Industry

http://stats.bls.gov/news.release/anpay2.toc.htm
Document: Annual Pay Levels in Metropolitan Areas

Women's suffrage
http://lcweb2.loc.gov/ammem/rbnawsahtml/nawstime.html
Document: One Hundred Years Towards Suffrage: An Overview

Women workers, characteristics of
gopher://english.hss.cmu.edu
Path: Gender/Facts on Working Women
Document: 20 Facts on Women Workers

Worker displacement
FTP://ftp.bls.gov/pub/news.release/disp.txt
Document: Worker Displacement During the Early 1990s

Workplace, fatalities/illnesses/injuries
http://ohr.systoc.com/docs/blsstats.htm
Document: Characteristics of Injuries and Illnesses Resulting in
Absences from Work, 1994

FTP://ftp.bls.gov/pub/news.release/osh.txt
Document: WorkPlace Injuries and Illnesses in 1994

http://ublib.buffalo.edu/libraries/e-resources/ebooks/
records/7162.html
Document: National Census of Fatal Occupational Injuries

Work, journey to
http://www.census.gov/ftp/pub/population/www/journey.html
Document: Census Bureau - Journey to Work

Work Stoppages, annual
http://stats.bls.gov/news.release/wkstp.toc.htm
Document: Major Work Stoppages, annual

World conflicts
http://www.emory.edu/CARTER_CENTER/demo.htm#conres
Document: Conflict Resolution Program